T0202375

Non-Ideal Epistemology

Non-Ideal Epistemology

ROBIN McKENNA

Great Clarendon Street, Oxford, OX2 6DP,
United Kingdom

Oxford University Press is a department of the University of Oxford.
It furthers the University's objective of excellence in research, scholarship,
and education by publishing worldwide. Oxford is a registered trade mark of
Oxford University Press in the UK and in certain other countries

© Robin McKenna 2023

The moral rights of the author have been asserted

Published in the United States of America by Oxford University Press
198 Madison Avenue, New York, NY 10016, United States of America

British Library Cataloguing in Publication Data
Data available

Library of Congress Control Number: 2023930546

ISBN 978–0–19–288882–2

DOI: 10.1093/oso/9780192888822.001.0001

Printed and bound by
CPI Group (UK) Ltd, Croydon, CR0 4YY

Links to third party websites are provided by Oxford in good faith and
for information only. Oxford disclaims any responsibility for the materials
contained in any third party website referenced in this work.

To Sophie, Edie, and Henry
It wouldn't be a family without them

Contents

Preface   ix

1. What Is Non-Ideal Epistemology?   1
   1.1 Three Aims   3
   1.2 Three Faces of Non-Ideal Epistemology   5
   1.3 Non-Ideal Epistemology and Feminist Epistemology   11
   1.4 Overview   15

2. Ideal and Non-Ideal Theory   19
   2.1 Ideal Theory in Ethics and Politics   22
   2.2 Full vs. Partial Compliance Theory   28
   2.3 Utopian vs. Realistic Theory   29
   2.4 Mills and Ideal Epistemology   31
   2.5 Objections   35

3. Anderson and Goldman on Identifying Experts   45
   3.1 Goldman on Identifying Experts   46
   3.2 Goldman and Ideal Theory   49
   3.3 Anderson on Identifying Experts   50
   3.4 Goldman vs. Anderson   56
   3.5 Two Tasks for Anderson   59

4. Persuasion and Paternalism   61
   4.1 Non-Ideal Institutional Epistemology   63
   4.2 Gathering the Evidence   66
   4.3 Epistemic Paternalism and Intellectual Autonomy   70
   4.4 Riley on Nudging and Epistemic Injustice   74
   4.5 Meehan on Nudging and Epistemic Vices   78
   4.6 Tsai on Rational Persuasion and Paternalism   81
   4.7 Intellectual Autonomy   85

5. Intellectual Autonomy   87
   5.1 What Is Intellectual Autonomy?   88
   5.2 Against Carter on Intellectual Autonomy   92
   5.3 More on Motivated Reasoning   96
   5.4 Against Roberts and Wood on Intellectual Autonomy   98
   5.5 Intellectual Autonomy and Epistemic Paternalism   100
   5.6 Becoming Intellectually Autonomous   102

6. The Obligation to Engage   105
   6.1 Mill on the Obligation to Engage   107
   6.2 Cassam on the Obligation to Engage   110
   6.3 The Obligation to Engage in Inhospitable Environments   112
   6.4 The Obligation to Engage and Epistemic Exclusion   116
   6.5 Full vs. Partial Compliance Theory   121
   6.6 Ballantyne, Fantl, and Srinivasan   124
   6.7 Towards a Non-Ideal Theory   129

7. Liberatory Virtue and Vice Epistemology   133
   7.1 Liberatory Virtue Epistemology   134
   7.2 Medina on Intellectual Virtue and Vice   138
   7.3 Making Sense of Responsibility   142
   7.4 Medina on Epistemic Responsibility   147
   7.5 Ideal Theory and Epistemic Responsibility   153

8. Scepticism Motivated   157
   8.1 Even More on Motivated Reasoning   158
   8.2 The Unreliability of Politically Motivated Reasoning   161
   8.3 Politically Motivated Reasoning and Basing   166
   8.4 Debunking Arguments   174
   8.5 Scepticism in Non-Ideal Epistemology   175

References   181
Index   195

Preface

This book has its origins in a class I taught on social epistemology at the University of Vienna in 2016. While preparing the class I was struck by a divide between two camps in social epistemology. One camp worked with idealized models of human beings, the social interactions between them, and the social spaces in which they interacted. This camp tended to focus on foundational issues in the epistemology of testimony (under what conditions am I justified in accepting what someone tells me?) and disagreement (what should I do when I learn that someone whom I regard as my epistemic peer disagrees with me?). While they acknowledged the importance of social interactions, this camp tended to view social interactions as a means of transferring epistemic goods (knowledge, information), and paid little attention to the ways in which social power differentials coloured and shaped these interactions.

The other camp did not do these things. When they talked about the same issues as the first camp, their interests seemed different. They were less concerned with the foundational issues and more concerned with the ways in which social power differentials shaped testimonial and other social interactions. More generally, they wanted to ask different questions. In what ways are our systems of knowledge production and dissemination dysfunctional? How might we improve these systems? How do our social identities and situations influence the evidence to which we have access? Can epistemological frameworks and systems themselves be sources of oppression? Why is it that so many of us are ignorant of the reality of oppression and injustice? What can epistemology do to help us answer all these questions?

In teaching the class, I used this divide as a way of framing the material for the students. The first camp did what I call 'ideal epistemology'. The second camp did what I call, for obvious reasons, 'non-ideal epistemology'. I suggested to my students that this divide was, in many ways, more useful than the divide between 'traditional' and 'social' epistemology. Whether or not they found this helpful, I have been thinking about this divide ever since. This book is my attempt to set out what I think about it. In the pages that follow I argue that the divide between ideal and non-ideal epistemology is as—if not more—important as the divide between traditional and social epistemology. However, this is not primarily a work in philosophical methodology. My basic aim is to make the case for non-ideal epistemology by doing it. I argue that some epistemological issues and problems call for a non-ideal rather than an ideal approach. This book is therefore a defence of non-ideal epistemology.

While this book is a defence of non-ideal epistemology, it is not necessarily, or at least not primarily, a critique of ideal epistemology. The part of the more general picture I have struggled the most with is the question of whether ideal and non-ideal epistemology are in opposition to each other. This question is much like the question of whether traditional and social epistemology are in opposition to each other. With traditional and social epistemology, a large part of the difference between them is just that they are often about different things. The same goes for ideal and non-ideal epistemology. Where the ideal epistemologist might want to know what knowledge is, the non-ideal epistemologist wants to know why it is that some people do not have it and what we can do about the fact they do not have it.

That said, my view is that ideal and non-ideal epistemology can come into conflict. Imagine we are interested in the question of whether we all have an obligation to engage with challenges to our beliefs. Many are inclined to answer in the affirmative, for reasons familiar from John Stuart Mill's famous discussion of freedom of expression and the value of robust critical debate in *On Liberty*. Mill's discussion of these matters might strike the reader as a little idealistic. One wants to agree that an idealized version of debate might have the sorts of benefits Mill alleges, but many—me included—are inclined to dispute whether actual debate is much like the idealized version of debate Mill seems to have in mind.

Now, my claim is not that the ideal epistemologist *must* give an affirmative answer to the question of whether we are under an obligation to engage with challenges. It may be possible to argue, from an ideal perspective, that we are under no such obligation. But the Millian picture of the value of robust critical debate is particularly attractive to someone who is inclined to think that, at least in its essential respects, actual debate is like the idealized version of it that Mill envisages. It is, however, not attractive to someone who is inclined to think that, at least with respect to certain issues (e.g. politically contentious issues), and for certain people (e.g. members of marginalized groups), actual debate is little like how Mill imagines it to be. As a result, we can expect the Millian picture to be more attractive from the standpoint of ideal epistemology than it is from the standpoint of non-ideal epistemology. If you think—as I do—that the Millian picture is wrong, then this points to one place where ideal and non-ideal epistemology can come into conflict. The pages that follow highlight some other places.

However, I want to highlight that, while ideal and non-ideal epistemology can come into conflict, they often do not. It may even be that they usually do not. In many ways, then, the situation with ideal and non-ideal epistemology parallels the situation with traditional and social epistemology. Traditional and social epistemology are two ways of doing epistemology that complement each other. A 'complete' epistemology needs both. This point is particularly important given the current state of epistemology. It is fair to say that non-ideal epistemology is increasingly popular. But ideal epistemology, whether in a traditional or more social form, dominates the syllabi of standard epistemology courses, and—at least until very recently—the pages of the major journals. If it really is true that ideal and non-ideal epistemology complement each other, and if they are—as I think—equally deserving of our time and attention, then this is an imbalance that needs to be corrected.

Several people have helped shape my thinking about these issues over the years. I would like to take this opportunity to thank them and apologize to anyone I have forgotten to mention.

The first group of people I want to thank are those who have read and commented on portions of the book at various stages of its development. Special thanks are due to Veli Mitova and to two anonymous readers for Oxford University Press. The book is much improved due to their many insightful comments and probing objections. Thanks also to Joshua Habgood-Coote for a helpful set of comments on Chapter 2, Nick Hughes for several objections to non-ideal epistemology that I work through in Chapter 2, the members of the Oxford Epistemology Group (especially Nick and Rachel Fraser) for their invaluable comments on Chapter 5, Daniella Meehan for her detailed comments on my discussion of her work Chapter 5, and to Miriam McCormick for urging me to discuss Jeremy Fantl's work on open-mindedness in Chapter 6. Finally, this book incorporates some previously published work (see below for details), on which I received helpful comments from Guy Axtell, Amiel Bernal, Jeroen de Ridder, Michael Hannon, Kirk Lougheed, Jon Matheson, and reviewers for the *Canadian Journal of Philosophy* and the *Journal of Applied Philosophy*.

The second group of people I would like to thank are those who gave me comments on presentations of material from the book. I presented a version of Chapter 2 to Joachim Horvath's philosophical methodology research group at the Ruhr-University Bochum. Thanks to Joachim for the invitation and a stimulating discussion. Thanks also to Steffen Koch for asking a question about ideal and non-ideal theory in political philosophy that helped me to reshape the chapter. I presented a version of Chapter 6 to the University of Glasgow's Epistemology Work in Progress Group, organized by the Cogito Research Centre. Thanks to J. Adam Carter for the invitation and to the audience for their questions. I cannot recall who said what, but the many objections were helpful in figuring out how to frame the chapter and the argument of the book. (I suspect the most forceful objections were from Mona Simion and Chris Kelp.) Finally, I presented versions of Chapter 7 at a workshop on Epistemic Blame at the University of Johannesburg, organized by Veli Mitova, and at the European Epistemology Network at the University of Glasgow, organized by the Cogito Research Centre. Thanks to both audiences for their helpful comments and to Veli for the invitation.

The third group of people I would like to thank are those who I have had conversations with about the book and related issues over the years. Let me single out Martin Kusch for special thanks here. Martin employed me for four years, but he also helped shape my thinking about epistemology and philosophy a lot more than might be apparent from reading this book. While I did not start drafting the book until after I left Vienna, once I started it quickly become clear that I had been writing whole parts of it in my head while I was in Vienna. I expect long conversations with Martin about epistemology and its tendency to idealize were the impetus. (I also expect he will complain: where is the sociology of scientific knowledge in all this?) In Vienna, I also benefitted from long conversations with Natalie Alana Ashton, Delia Belleri, David Bloor, Dirk Kindermann, Katherina Kinzel, Anne-Kathrin Koch, Carlos Núñez, Lydia Patton, Katherina Sodoma, Johannes Steizinger, Niels Wildschut, and Dan Zeman. Thanks also to (this is a very incomplete list) Kristoffer Ahlstrom-Vij, Cameron Boult, J. Adam Carter, Quassim Cassam, Katherine Furman, Mikkel Gerken, Sandy Goldberg, Michael Hannon, Allan Hazlett, Jonathan Jenkins Ichikawa, Ian James Kidd, Aidan McGlynn, Alessandra Tanesini, Vid Simoniti, and to the members of the Social Epistemology Network and Board-Certified Epistemologists Facebook groups.

Finally, this book makes use of previously published material, though with several changes, and sometimes to ends quite different to those originally intended. Chapter 4 incorporates two book chapters: 'Persuasion and Epistemic Paternalism', published in Guy Axtell and Amiel Bernal (eds), *Epistemic Paternalism: Conceptions, Justifications and Implications* (pp. 89–104), Lanham: Rowman & Littlefield, 2020, used with kind permission from Rowman & Littlefield, and 'Persuasion and Intellectual Autonomy', published in Kirk Lougheed and Jon Matheson (eds), *Epistemic Autonomy* (pp. 113–31), Abingdon: Routledge, 2021, used with kind permission from Taylor and Francis. Chapter 8 incorporates two journal articles: 'Irrelevant Cultural Influences on Belief', published in *Journal of Applied Philosophy* 36 (5), (2019) 755–68, used with kind permission from John Wiley and Sons, and 'Skepticism Motivated: On the Skeptical Import of Motivated Reasoning', published in *Canadian Journal of Philosophy* 50 (6), (2021) 702–18 (co-written with J. Adam Carter), used with kind permission from Cambridge University Press. I would like to express my gratitude to my co-author and these journals and publishers for allowing me to borrow from this previously published work.

1

What Is Non-Ideal Epistemology?

What is epistemology? If you have taken an epistemology class, you were likely told that epistemology is the 'theory of knowledge'. (If you have taught an epistemology class, you perhaps told your students it is the theory of knowledge.) So understood, epistemology is concerned with the nature of knowledge, which for many epistemologists really means the conditions under which someone knows something ('S knows that p if and only if . . .').

If your epistemology class (I am assuming you had an epistemology class) was about this, you might have had the nagging feeling that there was something odd about the whole thing. You were probably asked to consider what you would say about imagined scenarios where someone believed something to be true on what seemed to them like good evidence but, unbeknownst to them, something odd was going on. Perhaps our imagined individual—let us call them Smith—was in a field. Smith, you were told, sees something that looks a lot like a sheep. Surely, you thought, that means Smith *knows* there is a sheep in the field. But then your teacher told you the twist. The thing Smith is looking at is really a rock that looks like a sheep but behind the rock is an actual sheep obscured from view. What now? Does Smith know that there is a sheep in the field?

It may be that you figured out how epistemologists usually answered these questions. Maybe you even got good at concocting your own examples to test the plausibility of this-or-that theory of knowledge. On the other hand, it may be that you decided the whole thing was not for you. Either way, it was probably clear to you that, whatever epistemology is, it is an extremely abstract and theoretical enterprise. You might have asked what relevance it has to your life. You might have wondered whether it has any applications to 'real world' issues and problems.

If you were fortunate enough to have taken your introductory epistemology class recently, your teacher might have said that things are different now. Epistemologists do not just talk about lone individuals (or people called Smith). They also do not just talk about the theory of knowledge. One exciting development is that they now talk about social interactions. You might have been asked what you think about examples featuring two individuals—let us call them Smith and Jones—who tell each other things. Perhaps Smith tells Jones that the train leaves in ten minutes. Is Jones justified in believing what Smith has told him without checking whether Smith is trustworthy first? Maybe you were also asked to consider scenarios where Smith and Jones disagree with each other. Perhaps Smith says that the restaurant bill works out at £20 each whereas Jones says it is £21.

Non-Ideal Epistemology. Robin McKenna, Oxford University Press. © Robin McKenna 2023.
DOI: 10.1093/oso/9780192888822.003.0001

Should Smith and Jones redo their sums until they come to an agreement, or should they each conclude that the other has made an arithmetical mistake?

It may be that these developments satisfied you. You might have said that at least you can see why these questions *might* matter. We have all been in situations where we had to decide whether to believe what someone has told us and are not able to check whether they are trustworthy or not. We have also all had disagreements over the bill. Alternatively, it may be that these developments did not satisfy you at all. You might have objected that these sorts of scenarios are too simplified to be of any practical relevance. You might have got frustrated about the lack of detail. Why would Jones approach a stranger to ask when the train leaves when he could just use an app on his phone? How does Smith know that Jones is not just trying to trick him into paying more than he needs to? Why can't they just calculate their shares of the bill using an app on their phones?

Let us continue our story. Imagine that, despite your misgivings, you ended up doing more advanced courses in epistemology. Perhaps you even ended up doing a PhD on the subject and teach it yourself. If nothing else, this gave you the vocabulary to articulate the misgivings you had always had about epistemology. Your first exposure to epistemology was to what epistemologists now often derisively refer to as 'Gettierology', which is a central topic in 'traditional epistemology'. Gettierology—so called because it was sparked by Edmund Gettier's 1963 paper 'Is Justified True Belief Knowledge?'—is the project of trying to find necessary and sufficient conditions for a subject S to know some proposition p. It turns out that, at least nowadays, many professional epistemologists do not like Gettierology either.

You were then exposed to what is called 'social epistemology', which considers the epistemological implications of social interaction (Goldman and O'Connor 2021). However, as you now recognize, the problem with social epistemology is that you can consider the epistemological implications of social interaction while working with highly idealized pictures of what social interaction, and the creatures who do the interacting, are like. Any dissatisfaction with your initial exposure to social epistemology was really with these highly idealized models. Your complaint was that these idealized models do not tell us much about social interactions and so, despite its promise, social epistemology is of far less 'real world' relevance than it might initially appear to be.

Before this veers (too far) into autobiography, let me get to the point. This book is about the idealized models of human beings and the social interactions between them favoured by many social epistemologists. My central aim is to argue that serious problems can result from working with these idealized models. This book is therefore at least in part a defence of the bemused reaction of our imagined student (ok, me) to their first exposures to epistemology.

I say 'in part' because it is not a complete defence. Frustratingly for our imagined student, I am not going to come down entirely on their side. My basic claim is that what we can call 'ideal epistemology'—the kind of epistemology that works with these idealized models—goes wrong in that we sometimes need to work with less idealized models of human beings and of the social interactions between them. There is, therefore, a need for what we can call 'non-ideal epistemology'. Where ideal epistemology works with idealized models of humans and social interaction, non-ideal epistemology works with less idealized, more realistic models.

But let me be clear. I do not claim that ideal epistemology is fundamentally misguided, or that it is always a mistake to work with idealized models of epistemic agents and the interactions between them. Idealization can be a valuable tool, and it would be a grave mistake to object to idealization per se, both in general and in epistemology. Two sorts of problems can, however, arise when you engage in idealization. The first is that you might end up ignoring phenomena that are of real interest because you work at a level of idealization from which they are rendered invisible. The second problem arises when you forget that idealization is a tool or—still worse—when you forget that you are engaging in idealization in the first place.

1.1 Three Aims

Let me get into the specifics. In this book, I set out to achieve three things. First, I demonstrate the importance of distinguishing between ideal and non-ideal epistemology. The main obstacle to recognizing the importance of this distinction is the existing distinction between traditional and social epistemology. What, you might ask, does the distinction between ideal and non-ideal epistemology give us that we do not already get from the distinction between traditional and social epistemology? My answer is that it is possible—indeed common—to pursue social epistemology in a way that is highly idealized. In the following chapters I argue that problems can result when we pursue social epistemology in an overly idealized fashion.

My claim is not that the distinction between ideal and non-ideal epistemology should *replace* the distinction between traditional and social epistemology. My claim is that we need to *complicate* this distinction by recognizing the differences between ideal and non-ideal approaches to social epistemology. More generally, my contention is that we should pause to consider what ties together the various approaches adopted, and issues considered, within the umbrella of 'social epistemology'. What does the debate between reductionists and anti-reductionists in the epistemology of testimony have in common with the literature on epistemic injustice and oppression? What does the debate between conciliationism and the steadfast view in the epistemology of disagreement have in common with the (epistemological) literature on group polarization? To say that they are all concerned with the epistemological implications of social interaction is fine, but it ignores the many differences. These differences have to do with the fact that some of these debates are predicated on certain idealizations while others are not.

Second, I show that, at least with respect to some issues and problems of (social) epistemological interest, the non-ideal epistemologist's approach is preferable to the ideal epistemologist's approach. I have already said this earlier, but it bears emphasis, so I will say it again. I do not argue that non-ideal epistemology is 'better' than Ideal epistemology (whatever that might mean), or even that it is usually the right approach to adopt. My claim is just that it is sometimes the right approach to adopt.

This claim is, I think, modest. I do not view this book as arguing for a particularly radical claim. However, there is one sense in which it is a little less modest. I suspect—though have no evidence to back this up—that many epistemologists are tempted by the view that ideal and non-ideal epistemology are about different things and so cannot come into conflict in the way I think they can. I also suspect that this attitude is common when it comes to feminist epistemology (for more on the difference between non-ideal and feminist epistemology, see §1.3). Feminist epistemologists have identified some important questions and brought new issues to the foreground of epistemology (for example, epistemic injustice). But—you might think—the key contribution of feminist epistemology lies in the ways in which it expands the field of epistemology, not in the ways in which it critiques central assumptions of traditional (and much social) epistemology.

Whether this attitude is common or not, I think it is misplaced (Ashton and McKenna 2020; Dotson 2014, 2018; Toole 2019, 2022). In the chapters that follow, I identify some places where ideal and non-ideal epistemology can—indeed, do—come into conflict. I therefore do not just intend to show that ideal epistemology must be *supplemented* by non-ideal epistemology. I intend to show that, at least in certain cases, it must be *replaced* by non-ideal epistemology.

Third, I make progress in several debates in social epistemology by adopting a non-ideal approach. In the chapters that follow, I defend the following claims, all of which I take to be characteristic of non-ideal epistemology (I say more about what ties these claims together in §1.2):

(1) Solving the problem of public ignorance about consequential political and scientific issues like global warming requires creating a better epistemic environment and better social institutions (Chapter 3, Chapter 4).

(2) An 'epistemic' form of paternalism is (sometimes) justified (Chapter 4, Chapter 5).

(3) We often should not strive to be intellectually autonomous (Chapter 5).

(4) Some (e.g. John Stuart Mill) think we should all engage with challenges to our views. But, *contra* Mill, some of us are under no obligation to engage with (certain) challenges to our views. Indeed, sometimes we can dismiss

challenges without engaging at all. More generally, our obligations as inquirers depend on and vary with aspects of our social situation such as our social roles and identities (Chapter 6, Chapter 7).

(5) The core idea behind 'responsibilist' virtue epistemology is that epistemically responsible agency involves manifesting the intellectual virtues and avoiding intellectual vice. But making good on this idea requires a contextualized and socialized conception of epistemic agency and responsibility (Chapter 7).

(6) If we take the empirical literature on political cognition seriously, we are pushed towards a form of scepticism about whether our beliefs about political and politically relevant scientific issues are justified (Chapter 8).

This book is emphatically not just an exercise in epistemological methodology. Indeed, I spend far more time on first-order epistemological issues and questions than I do on methodology. While I offer some broad methodological considerations (particularly in Chapter 2), my aim is primarily to make the case for non-ideal epistemology by *doing it* rather than by talking about doing it. In the process, I hope that the distinctive features of non-ideal epistemology will become clearer.

It might, however, be helpful if I say a little more at the outset about what non-ideal epistemology is and what I take its distinctive features to be. I also want to say something about how my approach in this book differs from other approaches, especially feminist epistemology, and provide the reader with an overview of the book's contents. I start with the task of saying what non-ideal epistemology is in §1.2, before turning to situating my approach with respect to other approaches in §1.3 and the overview in §1.4.

1.2 Three Faces of Non-Ideal Epistemology

My distinction between ideal and non-ideal epistemology is based on Charles Mills' work on ideal and non-ideal theory in ethics and political philosophy (Mills 2005, 2007). Following Mills, I understand ideal epistemology as an approach to epistemological issues and questions that involves certain characteristic idealizations. Non-ideal epistemology is, then, an approach to epistemological issues and questions that eschews these sorts of idealizations. The idealizations that are characteristic of ideal epistemology include:

- Idealizations about the nature and psychology of epistemic agents or inquirers (e.g. about their cognitive capacities).
- Idealizations about the interactions between inquirers (e.g. about the extent to which they are influenced by social power differentials).

- Idealizations about social institutions (e.g. about their capacity to produce and disseminate knowledge).
- Idealizations about the environments in which inquirers are embedded (e.g. about the prevalence of information over misinformation).

I expand on these brief remarks in Chapter 2 but for now let me say two further things. First, it may often be the case that ideal and non-ideal epistemology deal with different issues and questions. This a consequence of the fact that there is a connection between the sorts of idealizations we engage in and the issues and questions that seem most pressing. If you work with a highly idealized picture of epistemic agents on which you abstract away from aspects of their social situation such as their social identity or role, you are hardly going to consider whether there are interesting epistemological differences between differently situated agents (e.g. do they have access to different bodies of evidence, as feminist standpoint theorists suggest?). If you ignore the fact that there are power differentials between epistemic agents, you are hardly going to consider the epistemological consequences of social power differentials between epistemic agents (e.g. do we afford more credibility to agents with more social power?). What this tells us is that one way in which non-ideal epistemology might improve on ideal epistemology is by identifying issues and questions that are of epistemological interest but obscured by the idealizations that are typical of ideal epistemology.

Second, the reader might find it uninformative to be told that ideal epistemology is an approach to epistemology that deals in idealizations while non-ideal epistemology is an approach that avoids idealizations. But this is no less informative than standard ways of distinguishing between traditional (or individual) epistemology and social epistemology. On one way of drawing the distinction, traditional and social epistemology differ in that traditional epistemology focuses on socially isolated individuals while social epistemology focuses on individuals embedded in a social context. This is all very well, but it is not terribly informative. If you want to better understand the distinction between traditional and social epistemology, you are best advised to look at concrete examples of social epistemological projects and compare them with concrete examples of more traditional epistemological projects. Similarly, if you want to better understand the distinction between ideal and non-ideal epistemology, you need to look at concrete examples of non-ideal epistemology. The chapters that follow supply several such examples.

Over and above offering a characterization of ideal and non-ideal epistemology, this book identifies three key aspects or 'faces' of non-ideal epistemology. The first key aspect or face of non-ideal epistemology is a focus on systems of knowledge production and the social institutions that play a crucial role in these systems. Now, a focus on systems and institutions is of course also a key aspect of social epistemology. Consider, for example, what Alvin Goldman (2010a) calls 'systems-oriented social epistemology' or what Elizabeth Anderson (2006) calls 'institutional epistemology'. Because these approaches are quite similar, I will confine my attention to Anderson's institutional epistemology. For Anderson, institutional epistemology is a branch of social epistemology that looks at the epistemic powers of social institutions. It considers questions such as:

- What sorts of knowledge reside within our social institutions, and what sorts of problems is this knowledge needed to solve?
- What problems should we assign to these institutions?
- How can these institutions be (re)designed to improve their epistemic powers?

These questions can be answered, and the programme of institutional epistemology can be pursued, in an idealized fashion or a non-idealized fashion. To make things more concrete, let us focus on the third question. One way of answering it—a way typical of an ideal approach to institutional epistemology—would be to consider what the optimal (epistemic) design of a social institution like science might be. On this approach, a central question would be something like 'how might we design science as a social institution so that it produces knowledge of the things we want to know about and tackles the sorts of problems we want it to tackle?' Philip Kitcher's work on science is a prime example of this sort of approach to institutional epistemology (Kitcher 2001, 2011).

Another way of answering the third question—a way typical of non-ideal epistemology—would be to start with the social institution as it currently is and ask which concrete steps we could take to improve its epistemic design. The proposed modifications would need to be *evidence-based*. That is, there would need to be evidence that the proposed modifications would secure the desired epistemic improvement. In the first half of this book, particularly in Chapters 3 and 4, I pursue this non-ideal approach to institutional epistemology. I do this in the context of a pressing social and political problem: science denialism, in particular the various forms of global warming scepticism. My task will be to survey the literatures on the causes and psychological drivers of global warming scepticism and on effective strategies for persuading sceptics to change their mind. Based on a survey of these literatures, I will make concrete proposals for what science communicators can do to combat global warming scepticism. The first half of this book therefore illustrates the institutional face of non-ideal epistemology.

While non-ideal epistemology pays particular attention to social institutions, it does not ignore individuals. The second face of non-ideal epistemology is a view of epistemic agents or inquirers as deeply socially situated. A view of epistemic agents as in some sense socially situated is of course also a key aspect of social epistemology. But the crucial question is which aspects of social situation are viewed as of epistemological importance. It is one thing to acknowledge that epistemic agents are socially situated in the minimal sense that they depend on each other for information. It is quite another to hold that aspects of social situation such as social identity and role are relevant to our epistemic obligations and responsibilities. It is the second, deeper sense of social situatedness that is characteristic of non-ideal epistemology.

This point can be amplified by considering the literature on feminist epistemology (for more on feminist epistemology, see §1.3). One of the core ideas of feminist epistemologies is that epistemic agents are socially situated in precisely this deeper sense. It is striking that, while many of the classic texts in feminist epistemology take issue with traditional epistemology for ignoring our 'social situatedness', the same critique can be levelled at much social epistemology too (see Mills 2007). Take, for example, Lorraine Code's (1991) classic *What Can She Know?* Code's explicit target is what she calls 'S knows that p epistemologies': theories of knowledge that take the form 'S knows that p iff...' where 'S' stands for all knowers. Code thinks these 'S knows that p epistemologies' are mistaken because they ignore the fact that differently socially situated knowers have access to different bodies and types of evidence, have different cognitive capacities, and have different epistemic obligations and responsibilities.

If this is a good criticism of 'S knows that p epistemologies', it is also a good criticism of much of the literature on the epistemology of testimony, the epistemology of disagreement, or indeed many of the central topics in contemporary social epistemology. If we need to consider social situation, we cannot ask what the rational response to testimony or disagreement is without specifying the social situations of the relevant parties (does one occupy a more socially powerful position than the other does?). Code's objection is therefore not just an objection to 'S knows that p epistemologies'. It is an objection to an approach to epistemology that abstracts away from aspects of our social situatedness that may be epistemologically relevant. It is an objection to ideal epistemology.

In the second half of this book, particularly Chapters 6 and 7, I develop this objection. I start (in Chapter 6) by looking at John Stuart Mill's famous argument in *On Liberty* that we all have an obligation to engage with challenges to our beliefs. I argue that we can explain both why many find Mill's argument attractive and why it is mistaken if we understand his argument as driven by certain idealizations that are typical of ideal epistemology. Further, I argue that, once we abandon these idealizations, a different picture of our epistemic obligations and responsibilities emerges. On this picture, the nature and extent of your epistemic obligations and responsibilities depend on whether you can expect other inquirers to satisfy their epistemic obligations to you. Inquirers who cannot expect other inquirers to satisfy their obligations towards them may have different obligations than inquirers who can expect these things.

I then (in Chapter 7) develop the view that our epistemic obligations and responsibilities are socially situated using the framework of 'liberatory' virtue and vice epistemology.[1] In the process I address a (if not the) central challenge for liberatory virtue (and vice) epistemology. The liberatory virtue epistemologist emphasizes the extent to which our characters are the product of our social situation. But (you might think) a character trait only qualifies as a virtue (or vice) if the possessor is responsible for having it, and (you might also think) if our characters are the product of our social situation then we cannot be responsible for them. I argue that the version of liberatory virtue epistemology defended by José Medina in his 2012 book *The Epistemology of Resistance* has the resources to deal with this challenge. In the process, I reconstruct Medina's account of epistemic responsibility, which is distinctive in that it socially situates epistemic agency and responsibility. It is, therefore, a non-ideal account of epistemic responsibility.

The third face of non-ideal epistemology runs through the book and is more an argumentative strategy than a claim or thesis. Several of the chapters that follow develop a general objection to ideal epistemology. Rather than stating the objection in the abstract, let me give a compelling example of it. (I do not take up this example in the book; I have nothing to add to what has been said about it.)

One strand running through feminist epistemology and the philosophy of science is an objection to a common way of thinking about objectivity (Anderson 1995, 2017; Harding 1995; Longino 1997). On this way of thinking, being objective is a matter of being detached, disinterested, and not emotionally invested in the outcome of your inquiries. An inquiry is then objective to the extent that the inquirers are objective: they keep their personal views, feelings, and value commitments out of their inquiries. This way of thinking about objectivity expresses an intellectual ideal that I will call 'objectivity as detachment'. This ideal was—and in some circles still is—common in the philosophy of science. The thought is that what distinguishes scientific from other forms of inquiry is its commitment to the ideal of objectivity as detachment (Lacey 1999).

The objection to this way of thinking of objectivity developed by feminist philosophers of science combines two claims. The first, which is descriptive, is that the ideal of objectivity as detachment is unattainable. It is extremely hard—perhaps impossible—to genuinely not be emotionally invested in the outcome of your inquiries, or to keep your personal views and values out of it. Worse,

[1] I use the term 'liberatory epistemology' to cover approaches to epistemology that are methodologically similar to feminist epistemology but focused on aspects of social identity beyond gender. Charles Mills' work will be particularly important in what follows (especially Mills 2007), as will José Medina's (especially Medina 2012).

we are often blind to the fact that we are emotionally invested in the outcome of our inquiries, and to the ways in which our views and values influence their direction. This is the point of several influential works of feminist science criticism that emphasize the extent to which scientists were blind to the impact of sexist assumptions and biases on their work (see e.g. Fine 2010 and Keller 1985).

The second claim, which is normative, is that scientific inquiry is not improved by trying to approximate the ideal of objectivity as detachment. You might hold that, even though it is not possible to remain completely detached, or to keep all your personal views and values out of inquiry, the ideal of objectivity as detachment can still function as a regulative ideal. It is an ideal that we should try to approximate even though we are unlikely to reach it.[2]

The feminist rejoinder is that we should not try to approximate the ideal of objectivity as detachment because in doing so we are liable to worsen rather than improve our inquiries. The thought is that we improve inquiry—make it more likely to produce knowledge—not by trying to keep as many values out of it as we can but by making sure it is informed by the *right* values. Many works of feminist science criticism emphasize that, in cases where feminist political values guided scientific interventions, the result was better science. For instance, in her work Elizabeth Anderson has argued that social scientific research is better if it is informed by feminist political values (e.g. Anderson 2004).

The following chapters develop an objection to ideal epistemology that has a similar structure to this objection. That is, they argue that the ideal epistemologist proposes intellectual ideals and norms of inquiry that are not only unattainable (the descriptive claim) but also such that, in trying to attain them, we run the risk of doing worse epistemically than we would if we did not try to attain them (the normative claim). The claim is therefore that the ideals and norms proposed by the ideal epistemologist are often not regulative ideals. We should not try to approximate them at all. For example, in Chapter 5 I argue that, in striving to be intellectually autonomous, we run the risk of sacrificing other, more valuable, intellectual goals. The lesson is that it is often better not to strive to be intellectually autonomous because doing so will often worsen rather than improve our epistemic situation.

This should suffice for now as an explanation of what I mean by ideal and non-ideal epistemology and of the key claims I make in this book. I now want to turn to situating my project with respect to three others that might seem similar to my own.

---

[2] For interesting discussions of regulative ideals, see Emmet (1994) and Rescher (1987). Note that neither Emmet nor Rescher suggests that this way of thinking about objectivity is a regulative ideal. Their point is that there can be regulative ideals—ideals that are unattainable but still serve to regulate our practices. I agree that there can be regulative ideals but in the following chapters I try to show that many of the ideals proposed by ideal epistemologists are not regulative ideals.

1.3 Non-Ideal Epistemology and Feminist Epistemology

The first project I want to briefly situate my own with respect to is the sociology of scientific knowledge, particularly as developed by the 'Strong Programme'. Key figures in the Strong Programme include Barry Barnes, David Bloor, and Harry Collins (see Barnes 1977; Barnes and Bloor 1982; Bloor 1976; Collins 1985). The Strong Programme is not centrally concerned with epistemology or idealization within it (the focus is more on philosophy of science). But it is certainly true that its proponents view traditional epistemology as one of its targets, and part of the reason why it is among the targets is that traditionally epistemologists were not interested in the social causes of belief (Kusch 2010).

One of the central principles of the Strong Programme is the 'symmetry principle', which says that we should ask about the causes of belief without any regard to whether the beliefs in question are true or false, or rational or irrational (Barnes and Bloor 1982). If we do this, and foreground sociological and psychological factors in our explanations of beliefs, then the sorts of factors epistemologists like to cite (reasons, evidence) just will not feature.

Many epistemologists view the symmetry principle and the broader methodology it embodies with suspicion because they seem to amount to an attempt to debunk the authority of knowledge by reducing it to social interests and power (Boghossian 2006). Whatever the merits of this criticism, the proponents of the Strong Programme are clearly not concerned with the normative assessment of belief or processes of belief-formation. If we are not going to assess beliefs for truth or falsity, we are not going to be engaging in normative assessment of beliefs or believers. We are going to focus on the descriptive question of why people believe what they do, or why they work with the epistemic norms that they do, rather than the normative question of what they should believe or how they should conduct their inquiries. My interest in this book is in both the descriptive *and* the normative questions. While I think that any serious attempt to answer the normative question needs to start with the descriptive question (see Chapter 4), I do not think that the descriptive question replaces the normative question or makes it obsolete. My concerns are therefore different from those of 'Strong Programmers' and closer to those of traditional epistemologists.

The second project I want to situate my own with respect to is really another way of drawing the distinction between ideal and non-ideal epistemology. My initial gloss of the distinction was that, where ideal epistemology involves certain idealizations, non-ideal epistemology avoids these idealizations. But, as I will go on to detail in Chapter 2, there are many ways in which you might idealize. I highlight idealizations about human cognitive capacities, social institutions, social interactions, and the social environment. Another way of drawing the distinction between ideal and non-ideal epistemology focuses primarily on idealizations concerning our cognitive capacities (Carr forthcoming). On this way of drawing the distinction, ideal epistemologists are concerned with what perfectly rational, cognitively unlimited agents would believe while non-ideal epistemologists are concerned with the norms governing humans with their many cognitive limitations.

I have no objection to this way of drawing the distinction. Distinctions can be drawn in many ways. But you can always ask, of a particular way of drawing a distinction, whether it highlights or obscures the things you are interested in. This way of drawing the distinction (in terms of cognitive capacities) obscures the fact that you can recognize our cognitive limitations while still making all sorts of *other* idealizations. For instance, you might acknowledge our cognitive limitations yet still work with idealized pictures of the social interactions between cognitively limited agents, and about the environment in which they do the interacting. Because my aim is to identify an idealizing tendency in *social* epistemology, I need a way of distinguishing between ideal and non-ideal epistemology that goes beyond our cognitive capacities. Consequently, in the next chapter I foreground idealizations about social interactions, social institutions, and our social environment as well as idealizations about our cognitive capacities.

The result is a conception of non-ideal epistemology that is explicitly ethical and political in that it brings phenomena like injustice and oppression into the purview of epistemology. I do not claim that no other conceptions are available. But I do claim that my conception ties epistemology up with debates in social and political philosophy. Given that deeper engagement with social and political philosophy is a clear trend in twenty-first-century epistemology, my conception of non-ideal epistemology is well-placed to make sense of—and contribute to—these developments.

The third and final project I want to situate mine with respect to is feminist epistemology. Because there are many similarities, I will go into some detail. As what I said earlier should already make clear, in this book I make use of several insights from the feminist (and liberatory) epistemological literature. In places, I also try to contribute to this literature (e.g. Chapter 7, where I address a key challenge for liberatory virtue epistemology). It is, however, important to empha size that my distinction between non-ideal and ideal approaches to epistemology is not equivalent to the distinction between feminist and traditional or mainstream epistemology. This is for two reasons.

The first reason is that, in some respects, my distinction is *broader* than that between feminist and traditional or 'non-feminist' epistemology. Recall that non-ideal epistemology has three aspects or faces: a focus on institutions, a view of inquirers as deeply socially situated, and a distinctive argumentative strategy. Of these faces, only the second is particularly characteristic of feminist epistemology. Non-ideal institutional epistemology *can* be specifically feminist, as when social institutions are analysed through a 'feminist lens', or when an epistemological analysis of institutions serves feminist political goals. But in the first half of the book (Chapters 3, 4, and 5) I tackle the problems raised by 'science denialism', particularly global warming denialism, without adopting a specifically feminist approach to them.

The second reason is that, in other respects, my concerns are *narrower* than the concerns of feminist epistemologists. One important strand in feminist epistemology is what we might call the 'feminist theory of knowledge'; the feminist alternative to Code's 'S knows that p epistemologies'. In previously published work I have argued that the core feminist critique of 'S knows that p epistemologies' is best understood as methodological rather than as allied to any substantive claims about the nature of knowledge (Ashton and McKenna 2020; McKenna 2020).

Let me briefly rehearse this argument. The epistemological tradition treats certain kinds of knowledge—those amenable to the 'S knows that p' analysis—as paradigmatic. Feminist epistemologists invite us to consider what a theory of knowledge would look like if it were constructed around different paradigms. As Anderson puts it in her excellent overview of feminist epistemology:

> Mainstream epistemology takes as paradigms of knowledge simple propositional knowledge about matters in principle equally accessible to anyone with basic cognitive and sensory apparatus: "$2+2=4$"; "grass is green"; "water quenches thirst." Feminist epistemology does not claim that such knowledge is gendered. Paying attention to gender-situated knowledge enables questions to be addressed that are difficult to frame in epistemologies that assume that gender and other social situations of the knower are irrelevant to knowledge. Are certain perspectives epistemically privileged? Can a more objective perspective be constructed from differently gendered perspectives? (Anderson 2017)

This is a critique of a kind of ideal epistemology that takes the paradigm cases of knowledge to be the simple ones—the ones where social situation seems irrelevant. In contrast, the feminist theorist of knowledge takes the paradigm cases of knowledge to be the 'messy' ones—gendered and more broadly socially situated forms of knowledge. Anderson's point is that your choice of paradigms will inform your theory of knowledge. If we view simple cases as the paradigm, we will construct a theory of knowledge designed to accommodate simple cases and then try to extend it to messier ones. (Most likely, we will get stuck on the simple cases and never get to the messy ones.) If we view messier cases as the paradigm, we will end up with a messy theory of knowledge, or even no 'theory' of knowledge at all.

While I am sympathetic to this critique, neither it nor the kind of ideal epistemology it targets is my focus in this book. More generally, my target in this book is not the epistemologist who is interested in the theory of knowledge, the theory of justification, norms of belief and assertion, or anything of that sort. My target is rather the epistemologist who is engaged in an overly idealized form of what Quassim Cassam (2016) calls 'inquiry epistemology'. The 'inquiry epistemologist' is interested in the activity of inquiry—in how we go about extending our knowledge. They ask questions about the mechanics of inquiry (how do we inquire?). They ask normative questions about inquiry (what is a responsible inquiry like?) and inquirers (what is a responsible inquirer like?). But they are also interested in *improving* inquiry. They want to make it more effective (better at reaching truth) and more responsible (better at reaching truth in the right kind of way).[3]

However, it is possible to do inquiry epistemology in an overly idealized way. For example, the inquiry epistemologist might make use of idealizations about what inquirers are like, the environment in which they conduct their inquiries, how they interact with each other, or the social institutions which aide or hinder their inquiries. As a result, their proposals for improving inquiry might be such that trying to put them into practice would be likely to achieve the opposite result—to make us less likely to achieve our epistemic goals. In the following chapters, I argue that some inquiry epistemologists (including Cassam) fall into precisely this trap. In making this case I will often draw on work from feminist epistemology that, in my terms, urges the importance of adopting a non-ideal approach to inquiry epistemology (see Chapters 6 and 7).

You might object that what I am calling 'inquiry epistemology' has its proper home in ethics rather than epistemology. More modestly, you might worry that inquiry epistemology blurs the distinction between epistemology and ethics in a way that is confusing and perhaps even problematic. If the reader has these concerns, be forewarned. In the following chapters, I spend little time on whether what I am doing is best understood as epistemology, ethics, or some combination of them both. I also do not make much effort at keeping epistemological questions separate from ethical ones. But this lack of care is, I like to think, principled. I do not think we can neatly distinguish between epistemology and ethics, nor do I think that we should try. I am interested in norms of inquiry and if we need to blur the divide between epistemology and ethics to profitably think about them

---

[3] A few remarks about 'inquiry epistemology' that do not fit into the main body of the text. First, Cassam credits Alfano (2012) with coining the term and Hookway (2003) with articulating the idea in a particularly clear fashion. But the general approach is far from new. Indeed, it is arguably what some of the founding figures of epistemology as we know it today (Descartes, Locke) were up to (Pasnau 2017; Wolterstorff, 1996). Second, by 'inquiry epistemology' I mean something like what Nathan Ballantyne (2019) calls 'regulative epistemology'. I prefer 'inquiry epistemology' to 'regulative epistemology' because, while I applaud Ballantyne's aspiration to *improve* inquiry rather than just *describe* it, it is important not to downplay the importance of arriving at a proper understanding of how inquiry works (see §4.1). Third, the relationship between inquiry epistemology and the literature on epistemic normativity (norms of action, assertion, and belief) is a little tricky. I briefly touch on this in the next footnote, in §2.4, and again at the beginning of Chapter 6. For relevant discussion, see Friedman (2020).

then, as far as I am concerned, the divide can be blurred. I return to this issue (if only briefly) in Chapters 4 and 6.

1.4 Overview

Finally—and apologies for taking so long to get there—here is the overview. As already advertised, Chapter 2 ('Ideal and Non-Ideal Theory') develops my initial characterizations of ideal and non-ideal epistemology. I start by looking at the debate between ideal and non-ideal theory in ethics and political philosophy. I then consider whether we can construct epistemological analogues of the ways in which the distinction between ideal and non-ideal theory has been understood in this debate. I look at three ways in which the distinction has been drawn: John Rawls' distinction between full and partial compliance theory (Rawls 1971), a distinction Laura Valentini has drawn between utopian and realistic theory (Valentini 2012) and Mills' list of the sorts of idealizations he takes to be typical of ideal theory (Mills 2005). I argue that Mills' characterization is more useful than Rawls' because it is more general. Further, I argue that realistic theory in Valentini's sense encompasses both ideal and non-ideal epistemology and so does little to advance our understanding of the differences between them. I also explain why ideal and non-ideal epistemology are not necessarily opposed to each other, further situate my project with respect to social epistemology, and address several objections to my project.

Chapter 3 ('Anderson and Goldman on Identifying Experts') illustrates the contrast between ideal and non-ideal epistemology via a detailed case study. I focus on the problem of the identification of expertise and on the contrasting approaches taken to this problem by Alvin Goldman (2001) and Elizabeth Anderson (2011). I argue that Goldman's approach is a clear example of ideal epistemology while Anderson's is closer to non-ideal epistemology. I also argue that Anderson's approach to the problem is preferable to Goldman's. Where Goldman sees the problem as one that arises for individual inquirers, Anderson sees the problem in more institutional and political terms. She is interested in how we might construct better systems of knowledge production and a better epistemic environment. She is, therefore, pursuing a non-ideal form of institutional epistemology.

The two chapters that follow address two tasks that need to be carried out to properly defend a non-ideal institutional epistemology like Anderson's. The first is to say more about how we might construct better systems of knowledge production and dissemination and about how we might construct a better epistemic environment. Doing this requires looking in some detail at the empirical literature on how to 'market' science to secure the maximum possible public uptake. The second task is to deal with a fundamental objection to Anderson's approach, which is that what she proposes involves interfering with our intellectual autonomy (our right to make up our own minds on issues of importance to us).

Chapter 4 ('Persuasion and Paternalism') takes up both tasks. I start by identifying a set of 'science marketing' strategies that we have reason to believe would be effective in constructing a better epistemic environment—an environment in which individuals are more likely to form true beliefs about scientific issues, like global warming, which are relevant to public policy. I then argue that we need to take seriously the worry that, even if these strategies are effective, they infringe on our intellectual autonomy. I try to defuse this worry by distinguishing between interventions on our inquiries that bypass our critical faculties but are trying to enhance our intellectual autonomy and interventions that undermine our very capacity for autonomous thinking and deliberation. The sorts of 'marketing strategies' I discuss in this chapter fall into the former category and so need not infringe on intellectual autonomy.

Chapter 5 ('Intellectual Autonomy') challenges the assumption that intellectual autonomy is valuable in the first place. My claim is that intellectual autonomy is an intellectual ideal or goal that many of us frequently and predictably fall short of. Moreover, it is an intellectual goal that often frustrates our other goals, in particular the goal of arriving at true beliefs about matters of importance to us. The chapter concludes that, in the absence of reasons for thinking that intellectual autonomy is more important than other intellectual goals, we often do better not to strive for it. The chapter finishes by addressing the implications for the argument of the previous chapter. If intellectual autonomy is less important than many suppose, then it is unclear why the fact (if it is a fact) that some form of intervention with someone's inquiries interferes with their intellectual autonomy is even a prima facie reason against that intervention. This chapter therefore amplifies the argument of the previous chapter by undercutting the original objection against 'science marketing'.

Chapters 3, 4, and 5 focus on the first, institutional, aspect or face of non-ideal epistemology. Chapters 6 and 7 turn to the second aspect or face, which is a view of epistemic agents or inquirers as deeply socially situated and of their epistemic obligations and responsibilities as depending on aspects of their social situation Chapter 6 ('The Obligation to Engage') focuses on one obligation in particular: the (supposed) obligation to engage with challenges to our beliefs. I start by outlining the best-known defence of the view that we have such an obligation, which is that given by John Stuart Mill in On Liberty. I then look at a rather different defence of the same view, which can be found in Quassim Cassam's (2019) book Vices of the Mind.

Where Mill's view is that we all have an obligation to engage with challenges to our beliefs because this is the best way of securing certain epistemic benefits, Cassam's view is that it is only by engaging with challenges to our beliefs that we earn the right to them. I argue, against both Mill and Cassam, that we do not all have this obligation, at least with respect to all our beliefs. As I argue, the problem with both Mill's and Cassam's arguments is that they are based on assumptions that are typical of ideal epistemology. These assumptions are: (i) that inquirers will satisfy their obligations, and (ii) that our epistemic environment is (relatively) hospitable.

Once we recognize that these assumptions are false, a different picture of our epistemic obligations and responsibilities emerges. On this picture, the nature and extent of your epistemic obligations and responsibilities depends on whether you can expect other inquirers to satisfy their epistemic obligations to you and on how hospitable the epistemic environment is for you. Inquirers who cannot expect other inquirers to satisfy their obligations towards them, or for whom the epistemic environment is not hospitable, may have different obligations and responsibilities than inquirers who can expect these things and for whom the epistemic environment is (relatively) hospitable.

Chapter 7 ('Liberatory Virtue and Vice Epistemology') develops the idea that our epistemic obligations and responsibilities depend on our social identities and situations using the framework of liberatory virtue (and vice) epistemology. Specifically, it develops this idea through the version of liberatory virtue epistemology defended by José Medina in his 2012 book *The Epistemology of Resistance*. I argue that Medina's virtue epistemology is distinctive in that it socially situates epistemic agency and responsibility. It is therefore a recognizably non-ideal form of virtue epistemology.

Chapter 8 is a little different from the earlier chapters. It considers the epistemological implications of the empirical literature on motivated reasoning, which I draw on in earlier chapters. As this literature documents, the way we think about issues that matter to us (especially political issues) is quite different from how the epistemologist might imagine it to be. We often engage in so-called 'motivated reasoning': we gather and assess evidence in ways that serve our goals. For example, when presented with evidence that conflicts with our goal of thinking well of ourselves (thinking of ourselves as healthy, kind, compassionate, etc.), we tend to find ways of discounting that evidence. When gathering evidence about an important issue (e.g. global warming), we tend to gather evidence in ways that support a stance on the issue that serves our political goals (e.g. looking for evidence that humans are not responsible as a way of preventing political action). I argue that this provides the basis for an empirically driven argument for a form of scepticism. Specifically, it supports the conclusion that many of us do not have justified beliefs about political issues and scientific issues that have become politically contentious, such as global warming.

I finish the chapter and the book by considering whether we can avoid this sceptical conclusion by making a move you might think is like the move I have made in earlier chapters. Can we argue that the fact (if I am right) that a kind of justification for our political and (some) scientific beliefs is unattainable provides a reason to revise our understanding of what having justification for these beliefs requires? I argue that we cannot because, unlike in the earlier chapters, it is not the case that striving for this kind of justification worsens our (or anyone else's) epistemic situation. This chapter therefore serves not only to illustrate the ways in which non-ideal epistemology might contribute to more traditional epistemological debates but also to clarify my more general objection to ideal epistemology. Ideal epistemology is problematic only to the extent that the goals and norms it proposes run the risk of worsening rather than improving our epistemic situation. The non-ideal epistemologist is not committed to the implausible claim that the mere fact that a goal is unattainable or a norm hard to follow is itself a reason to not try to attain or strive for it.

2

Ideal and Non-Ideal Theory

In political philosophy there is a heated debate about ideal and non-ideal theory. This debate is related to the wider debate about the influence of John Rawls on 'analytic' political philosophy.[1] On one side, some take Rawls to have reinvigorated analytic political philosophy and to have established a framework within which political problems and issues can be productively grappled with. On the other, critics of Rawls point to the distorting influence of his work in creating a discipline in which pressing political problems—like what to do about persistent racial injustice—are set to one side in favour of abstract theorizing about what a just society would look like. As Charles Mills (a prominent critic) puts it:

> Rawls himself said in the opening pages of "A Theory of Justice" that we had to start with ideal theory because it was necessary for properly doing the really important thing: non-ideal theory, including the "pressing and urgent matter" of remedying injustice. But what was originally supposed to have been merely a tool has become an end in itself; the presumed antechamber to the real hall of debate is now its main site.   (Yancy and Mills 2014)

I am not going to arbitrate the debate between Rawlsians and their critics. This book is about ideal and non-ideal theory in epistemology, not in political philosophy. Still, as we will see, looking at the debate in political philosophy helps us understand ideal and non-ideal theory in epistemology. My main aims in this chapter are to explain what ideal and non-ideal epistemology are, and to pinpoint the non-ideal epistemologist's core complaints about ideal epistemology.

The way in which I draw the distinction between ideal and non-ideal epistemology will be quite simple. Ideal epistemology is an approach to epistemology that involves making idealizations about the nature and psychology of epistemic agents or inquirers, the interactions between them, social institutions, and the epistemic environment in which they are embedded (compare

---

[1] I prefer 'analytic' to 'Anglo-American' because, while both labels are problematic, 'analytic' does not mischaracterize the tradition to which it refers as having its origins in English-speaking countries or carry the false implication that you can only find exponents of the tradition in English-speaking countries.

Non-Ideal Epistemology. Robin McKenna, Oxford University Press. © Robin McKenna 2023.
DOI: 10.1093/oso/9780192888822.003.0002

Mills 2005). Non-ideal epistemology is an approach that avoids making these idealizations.

My attempt to pinpoint the non-ideal epistemologist's core complaints about ideal epistemology will be a little more interesting. Following Mills, I argue that there are two main problems with ideal epistemology. First, in making these sorts of idealizations, the ideal epistemologist runs the risk of ignoring important phenomena. Second, and moreover, they run the risk of constructing an inadequate epistemology. The ideal epistemologist runs the risk of ignoring important phenomena because they abstract away from phenomena of epistemological interest. The ideal epistemologist runs the risk of constructing an inadequate epistemology because, in building in various idealizations, they often end up proposing intellectual goals and norms of inquiry that are not only unattainable but such that, in trying to attain or follow them, we are likely to worsen rather than improve our epistemic situation.

Before continuing, three comments. First, ideal and non-ideal epistemology are best viewed as *tendencies*. Ideal epistemologists tend to make certain idealizations while non-ideal epistemologists tend to avoid these idealizations. It is important to view these approaches this way because, while we can make sense of 'fully ideal' theory, it is hard to make much sense of 'fully non-ideal' theory. Even the non-ideal epistemologist engages in *some* idealization and abstraction for the simple reason that idealization and abstraction is a necessary part of theorizing.

The crucial questions concern *which* idealizations you engage in. There is a difference between idealizations that matter given the question or issue you are addressing and idealizations that do not. The non-ideal epistemologist's complaint is that the ideal epistemologist tends to make idealizations that they should not make given the questions or issues they are trying to address. Of course, whether this complaint is true cannot be decided in the abstract. We need to look at concrete issues and consider which idealizations are permissible and which are not. In this book, I hope to convince you that there are cases where the ideal epistemologist makes idealizations that matter because they distort their analysis of the issue in question.

Second, there need not be any essential opposition between ideal and non-ideal epistemology. We can view ideal and non-ideal epistemology as different methodologies. Sometimes, these methodologies complement each other. Other times, they might give different answers to the same question, or differ as to what the crucial question or issue is. The chapters that follow illustrate both these possibilities. In Chapters 3, 4, and 5 we will see that ideal and non-ideal epistemology take different approaches to the problem of what to do about widespread public ignorance. In Chapters 6 and 7 we will see that ideal and non-ideal epistemology work with different pictures of what our obligations and responsibilities as inquirers are, and of what grounds these obligations and responsibilities.

Third, within epistemology there is no real parallel to the debate about ideal theory in political philosophy.[2] Why? It is certainly not because nobody is worried about the abstract or idealized nature of contemporary analytic epistemology. Indeed, a common complaint about twentieth-century analytic epistemology is precisely that it was too abstract, too idealized, too disconnected from real-world concerns. The most plausible explanation is that there is no real debate because there was already a big debate in (late) twentieth-century epistemology about traditional and social epistemology. Most epistemologists, if pressed, would say that this debate was about much the same issues as debates about ideal theory in political philosophy.

For those unfamiliar with the debate about traditional and social epistemology, let me briefly sketch the contours of it. Traditionally, epistemology was very individualistic (though perhaps not as individualistic as twentieth-century histories of it tend to suggest). But in the 1970s, epistemologists started to consider the epistemic implications of social interactions. This new movement—called social epistemology—corrected the overly individualistic focus of traditional epistemology by investigating the epistemic effects and dimensions of social interactions and social systems (Goldman and O'Connor 2021). Social epistemology developed against the background of traditional epistemology, but it corrected traditional epistemology's rigid focus on the beliefs of socially isolated individuals.

It may be that this is the right *sociological* explanation why there is no parallel in epistemology to the debate about ideal theory in political philosophy. But I do not think it is a good *philosophical* rationale. One of my main contentions, which I will substantiate in this chapter, but also in this book, is that the debate between ideal and non-ideal epistemology is largely distinct from the debate between traditional and social epistemology. I say 'largely' because, while you can do social epistemology while still doing ideal epistemology, it is harder to see how you could do non-ideal epistemology without also doing social epistemology. (Hard but not impossible. Bortolotti 2020 is a case in point, as may be Cassam 2014 and Kornblith 2012.) Setting these complications aside, we can say that non-ideal epistemology is a *way* of doing social epistemology. But it is a way of doing social epistemology that is quite different from how social epistemology is sometimes practised. I contend that it is different enough that we need new labels to mark the distinction.

Here is the plan for the rest of the chapter. I start by identifying three ways in which the distinction between ideal and non-ideal theory has been drawn in political philosophy (§2.1). This section prepares the ground for the next three

---

[2] My claim is not that there is no debate at all. You could read Mills (2007) as trying to initiate a debate about ideal theory in epistemology, and a non-ideal perspective is implicit in a lot of the subsequent literature (see e.g. Medina 2012). More recent discussions of ideal and non-ideal theory in epistemology include Begby (2021, ch. 3) and Lackey (2018). Still, it is fair to say that there is far less debate in epistemology than in political philosophy.

sections, which go through three ways of distinguishing between ideal and non-ideal epistemology that parallel these ways of distinguishing between ideal and non-ideal theory (§§2.2–2.4). I argue that the best way of drawing the distinction is, following Mills, in terms of certain idealizations that are characteristic of ideal epistemology and the avoidance of which is characteristic of non-ideal epistemology. I also clarify what I take the non-ideal epistemologist's core complaints about ideal epistemology to be. My aims in this chapter are preparatory—I am trying to say what is at issue in the debate between ideal and non-ideal epistemology, not arguing for one side of the debate over the other. However, in places I anticipate the argument of the rest of the book. This is particularly the case in §2.5, where I deal with some objections to non-ideal epistemology and to my project as a whole.

2.1 Ideal Theory in Ethics and Politics

The debate about ideal and non-ideal theory in ethics and political philosophy is messy. There are several ways of drawing the distinction and of identifying the crucial points of contention. My aim in this section is not to provide an exhaustive overview of this debate, still less to take sides in it. My aim is just to highlight three ways in which the distinction between ideal and non-ideal theory has been drawn that will be useful when we turn to epistemology.

Let us start with how Rawls understood the distinction. In his *Theory of Justice* Rawls distinguishes between full and partial compliance theory. Rawls' concern in this book is with what justice demands of us. The distinction between full and partial compliance theory is a distinction between two parts of a full answer to this question:

> The first or ideal part [of a theory of justice] assumes strict compliance [with the principles of justice] and works out the principles that characterize a well-ordered society under favorable circumstances... [This] ideal part presents a conception of a just society that we are to achieve if we can. Existing institutions are to be judged in the light of this conception.   (Rawls 1971, p. 216)

Strict or full compliance theory assumes that (i) people will comply with the principles of justice and (ii) society is sufficiently well developed to realize justice. It then asks what justice demands of us in such a society. Much of *Theory of Justice* is devoted to answering this question. The answer to it gives us a conception of what we are working towards—what our efforts to achieve justice are trying to achieve.

The second or non-ideal part of Rawls' theory asks what justice demands of us in situations where one (or both) of these assumptions is false. That is, it asks what justice demands of us in situations where there is only partial compliance with the principles of justice and/or where society is not sufficiently well developed to realize justice. Because we are often in situations like this, partial compliance theory tells us what we should do *now*, in our non-ideal world, to make it more just. Importantly, what we should do now, in our non-ideal world, need not be the same as what agents who inhabited a world where there was full compliance with the demands of justice should do. (If it were the same, there would be no need for partial compliance theory as well as full compliance theory.) Let me try to make this clearer with a (somewhat simplistic) example.

Imagine you are interested in figuring out a just set of policies for combatting and mitigating global warming. Full compliance theory would ask what 'climate justice' demands of us on the assumptions that (i) everyone (individuals and states) will comply with these demands and (ii) climate justice is achievable. In contrast, partial compliance theory would consider what the demands of climate justice are given that either (i) many other actors will not comply with them (as seems likely) or (ii) climate justice is not attainable (as also seems likely). The crucial point is that these approaches may deliver different answers. Climate policies that would be just if they were imposed on and followed by everyone might not be just if they are imposed on everyone but only followed by some.

Let me now turn to a second way of drawing the distinction between ideal and non-ideal theory. In her excellent overview of the debate, Laura Valentini suggests we view the distinction as one between 'utopian' and 'realistic' theory:

> [W]e need to distinguish between 'fully utopian' theories, which altogether reject the need to place feasibility constraints on principles of justice, and 'realistic' theories, which accept some such constraints.   (Valentini 2012, pp. 656–7)

Valentini's (fully) utopian theorist views justice as akin to a Platonic ideal while her realistic theorist insists that the problem with Platonic justice is that we will never be able to realize it. Her realistic theorist wants instead to urge that we focus on a more realistic or attainable ideal. You might think this stacks the deck against the utopian theorist. The utopian theorist rejects *all* feasibility constraints while the realistic theorist only needs to recognize *some* such constraints. But the better way to put it is that utopian theories place minimal feasibility constraints on principles of justice while realistic theories place substantial constraints.

More generally, Valentini's thought is that the distinction between utopian and realistic theory tracks a distinction between two different projects you might have as a political philosopher. The first project is evaluative rather than normative. It asks what a just society would look like, not what we should do to make our society more just. The second project is normative. Given some conception of what a just society looks like, it asks what we should do to make our society more just. Valentini's suggestion is that utopian theory is appropriate for the first project but not the second while realistic theory is appropriate for the second but not the first:

> *If* we want a yardstick for measuring how much our society is failing compared to a fully ideal one, then we need to make minimal factual assumptions, such as moderate scarcity, limited altruism, and perhaps reasonable disagreement. That is, we must not include unjust human conduct. *If*, on the other hand, we wish to design prescriptions that are likely to be effective, *given* some common flaws in human behaviour, then we better factor in more real-world constraints.
>
> (Valentini 2012, p. 660)

For certain purposes, we may want a way of measuring how well our society is doing compared to an ideal society. For that, we need a theory that is based on minimal factual assumptions, that is, utopian theory. For other purposes, we want a way of deciding what we should do here and now. For that, we need a theory that builds in lots of factual assumptions, that is, realistic theory. If this is right then, not only is utopian theory more plausible than it might appear, but there is also less of a disagreement between it and realistic theory than there might initially appear to be.

This version of the distinction between ideal and non-ideal theory may best capture what is (and is not) at issue in the debate in ethics and political philosophy. However, whatever its merits in ethics and politics, an analogous distinction between utopian and realistic theory does little to advance our understanding of ideal and non-ideal theory in epistemology. I explain why in more detail in §2.3, but to anticipate, the problem is that the closest epistemological analogue of Valentini's realistic theory is what I call inquiry epistemology, but inquiry epistemology can be pursued in a more or less ideal way. What we really need, at least in epistemology, is a distinction between more and less ideal versions of realistic theory. If this is right, then distinguishing between utopian and realistic theory does little to advance our understanding of the difference between ideal and non-ideal epistemology.

Finally, let me turn to how Mills draws the distinction between ideal and non-ideal theory in his influential paper '"Ideal Theory" as Ideology'. Here is Mills:

> What distinguishes ideal theory is the reliance on idealization to the exclusion, or at least marginalization, of the actual ... this is *not* a necessary corollary of the operation of abstraction itself, since one can have abstractions ... without idealizing. But ideal theory either tacitly represents the actual as a simple deviation from the ideal, not worth theorizing in its own right, or claims that starting from the ideal is at least the best way of realizing it.   (Mills 2005, p. 168)

Mills recognizes that all theorizing involves an element of abstraction and idealization. To object to abstraction or idealization per se would just be to object to theorizing per se. What Mills objects to are the *kinds* of abstractions and idealizations that are characteristic of ideal theory. Ideal theory traffics in idealizations that 'represent the actual as a simple deviation from the ideal'. More specifically, he identifies four sorts of idealizations that tend to do this:

(1) Idealizations about agents: Ideal theory ignores, or at least downplays, certain characteristics of the agents with which it is concerned. For example, it ignores the fact that some of them have power over others.
(2) Idealizations in psychology: Ideal theory ignores, or at least downplays, our cognitive limitations.
(3) Idealizations about social institutions: Ideal theory ignores, or at least downplays, the reality that some social institutions are oppressive.
(4) Idealizations about our environment: Ideal theory ignores, or at least downplays, the influence of prevalent misconceptions and illusions on our beliefs and our conduct.

Non-ideal theory, then, is theorizing that eschews these sorts of idealizations. Of course, it is possible to avoid some of these idealizations while making others. Properly speaking, we need to view ideal and non-ideal theory as standing on a continuum. 'Ideal theory' refers to theory that is closer to the ideal end; 'non-ideal theory' refers to theory that is closer to the non-ideal end. I will not have much more to say about this point in what follows, but the reader should bear it in mind.

Why does Mills think ideal theory is problematic? He says this:

> If we start from what is presumably the uncontroversial premise that the ultimate point of ethics is to guide our actions and make ourselves better people and the world a better place, then the framework above [that of ideal theory] will not only be unhelpful, but will in certain respects be deeply *antithetical* to the proper goal of theoretical ethics as an enterprise. In modelling humans, human capacities, human interaction, human institutions, and human society on ideal-as-idealized-models, in never exploring how deeply different this is from ideal-as-descriptive-models, we are abstracting away from realities crucial to our comprehension of the actual workings of injustice in human interactions and social institutions, and thereby guaranteeing that the ideal-as-idealized-model will never be achieved.
>
> (Mills 2005, p. 170)

Mills is making two criticisms here that are worth highlighting and considering separately. His first criticism is that, if we build the sorts of idealizations that are typical of ideal theory into our theorizing, we will lack the tools to understand certain things. If we ignore the reality of oppression, human limitations, and the flawed nature of our environment and many social institutions, then we will lack the tools to understand and explain these things, still less the tools to do something about them. Mills' point here is that the myriad ways in which the actual world differs from an ideal world stand in need of explanation. It is hard, if not impossible, to see how ideal theory could provide this explanation, precisely because it obscures these features of the actual world. If this is right, then ideal theory cannot be a *complete* theory because it cannot explain everything that needs to be explained.

His second criticism is that the fact that, in an ideal world, we would behave in a certain way does not and cannot entail that, in our world, we should behave in that way. This point is related to what is known in economics as the 'problem of the second best' and has been dubbed 'the fallacy of approximation' by David Estlund (2019). Because Estlund's fallacy is easier to explain, I will focus on it.

Imagine a situation where you have a problem and the best (most ideal) solution to that problem is not available. You commit the fallacy of approximation when you infer, from the fact that the best solution to the problem is X, that the best thing you can do is whatever most closely approximates to X. For example, imagine I need to go shopping and the shops are far away. The best solution to my problem would be to get in a reliable and fully functional car and drive to the shops. Unfortunately, my car has dodgy brakes and is liable to break down at any moment. The best thing for me to do is not to get in my malfunctioning car and drive to the shops (I may well kill myself, or someone else). I should walk or get the bus.

While this example is a world away from political philosophy, the same point applies here too. Most would agree that, in an ideal world, everybody would be afforded equal opportunities regardless of their social identity or position. Even though we do not live in such a world, you might think it desirable to live as if we did and not factor social identity or position into any decisions you make. You might preach the importance of 'colour blindness' in evaluating job applications and institute policies forbidding affirmative action. While views about the rights and wrongs of affirmative action differ, it should be clear that the fact that there would be no need for affirmative action in an ideal world does not entail that there is no need for affirmative action in our non-ideal world. In an ideal world, there would be no need for affirmative action because the problem it is intended to solve does not exist. In our world, the problem it is intended to solve does exist.

More generally, behaving as if we live in an ideal world when we do not may lead to bad outcomes. Indeed, it may lead to worse outcomes than if we recognized some of the ways in which our world is not ideal and thought about how we could try to mitigate them. If this is right, then the problem with ideal theory is that it is *inadequate as a normative theory* because the normative prescriptions it yields need not serve our goals. Ideal theory sometimes needs to be *replaced* by non-ideal theory.

I said at the outset that I do not intend to arbitrate the debate about ideal and non-ideal theory in political philosophy. But let me make two comments that go a small way in this direction. The first concerns a possible misunderstanding of Mills. As I read him, Mills does not claim that the problem with ideal theory is that we are unlikely to be able to follow its normative prescriptions or realize the goals it sets for us. When Mills says that ideal theory does not provide us with guidance, he does not mean that it is too demanding. (It may be that it is too demanding, but that is not why it does not provide guidance.) He means that it provides us with *bad* guidance. In trying to follow the normative prescriptions set down by ideal theory, we run the risk of making things worse than we would have if we had not tried to follow or realize them.

We can illustrate these points via the example of colour blindness. For Mills, the problem with the ideology of colour blindness ('I don't see colour') is not that it is difficult to practise (though it may be). The problem with it is that, far from serving the ends it is intended to serve (racial justice), it ends up exacerbating racial injustice by hiding it. As we will see in the chapters that follow, the problem I raise for ideal epistemology is analogous. The ideal epistemologist holds that we have certain intellectual obligations and goals. It may be that these obligations and goals are so demanding that we cannot achieve or follow them. But that is not what is wrong with them. The problem is that they serve to frustrate our most basic intellectual goals, that is, having true beliefs or knowledge about the world around us. The problem is not that ideal epistemology 'sets the bar too high'. The problem is that trying to do what it tells us to do will often not serve to improve our epistemic position.

The second comment is that Mills' distinction between ideal and non-ideal theory subsumes Rawls' distinction between full and partial compliance theory. Full compliance theory embodies some of the idealizations Mills identifies as characteristic of ideal theory: it assumes a world where actors comply with their obligations (whatever those are) and an environment that is conducive to the realization of some goal we have (justice). It therefore involves idealizations about human agents and our environment. Partial compliance theory avoids these idealizations. (It is not surprising that Mills' distinction subsumes Rawls'; Rawls is one of the principal targets of Mills' criticisms of ideal theory.)

If this is right, then we should expect Mills' criticisms of ideal theory to apply to full compliance theory. Do they? The first criticism certainly seems to. Because it assumes full compliance with the demands of justice, full compliance theory obscures the fact that many do not comply with the demands of justice. You might think that the mere fact that many do not comply with the demands of justice calls out for explanation and full compliance theory clearly cannot supply the explanation because it assumes that everyone does comply.

Things are a bit less clear-cut with the second criticism. It certainly seems right to say that there is a difference between how we should act given the assumption that others will comply with their obligations (or that the environment is conducive to the realization of our goals) and how we should act given the recognition that others will not comply with their obligations. (Rawls recognizes this point—this is why we need partial compliance theory.) This is only a criticism if we combine full compliance theory with the further claim that we should—in our noncompliant world—act as we would in a fully compliant world. But why should we do that?

One way of seeing what Mills is getting at is to bear his first criticism in mind when considering his second. If he is right that ideal theory tends to ignore the many ways in which our world is far from ideal, then Ideal theorists are likely to misjudge the distance between our non-ideal (noncompliant) world and an ideal (compliant) world. As a result, they are likely to underestimate the extent to which the normative prescriptions of ideal theory are ill-suited to the actual world. Now, to repeat, my aim is not to arbitrate the debate between Rawlsians and their critics so I am not claiming that this can be made to stick as a criticism of Rawlsian full compliance theory. However, in the following chapters, I argue that something like this is a fair criticism of the epistemological analogue of full compliance theory and of ideal epistemology more generally. Ideal epistemologists tend to forget that their theories are built on idealizations and so to underestimate the extent to which their normative claims are ill-suited to our actual, non-ideal world.

Summing up, in this section I have identified three ways of distinguishing between ideal and non-ideal theory in ethics and politics. They are (i) full vs. partial compliance theory, (ii) utopian vs. realistic theory, and (iii) theorizing that makes certain idealizations vs. theorizing that eschews these idealizations. I suggested that the third way of drawing the distinction subsumes the first and indicated why the second way might not be that useful for my purposes. I now turn to the epistemological analogues of these distinctions.

2.2 Full vs. Partial Compliance Theory

We can start with the distinction between full and partial compliance theory. Rawls drew this distinction within the theory of justice, but a parallel distinction can be drawn within epistemology. Specifically, it can be drawn within what in Chapter 1 I called 'inquiry epistemology'. Inquiry epistemology is interested in the norms governing inquiry and the epistemic obligations and responsibilities of inquirers. These include norms and obligations governing the gathering of evidence, the identification of trustworthy sources, and how we should behave in the social interactions we enter into when conducting our inquiries.

One approach to inquiry epistemology, which is the analogue of full compliance theory, would be to ask how we ought to conduct our inquiries on the assumptions that (i) other inquirers will comply with their obligations and (ii) the environment is conducive to complying with these obligations. (I will say that, when the environment is conducive, it is 'epistemically hospitable'.) Another approach, which is the analogue of partial compliance theory, would be to ask how we ought to conduct our inquiries given that other inquirers may not comply with their obligations and that the environment may be epistemically inhospitable.

To see what this distinction amounts to in practice, let us consider an example, which I take up in more detail in Chapter 6. Imagine we want to figure out what to do when we meet another inquirer who disagrees with us. Should we engage with them and consider their objections in detail? If so, how much engagement is needed? If not, how do we know when not to engage?

Full compliance theory answers these questions based on two assumptions: other inquirers will comply with their obligations and the environment is epistemically hospitable. If we make these assumptions, we can make some further assumptions about what a dialogue between two inquirers with opposing views will be like. It will be a discussion between two well-informed people (they will have fulfilled their obligation to gather evidence). While they disagree—perhaps even fundamentally—they engage in discussion because they are committed to discovering what is true. Because they are committed to discovering what is true, they listen to what is said, advance arguments, and engage in good faith.

In contrast, partial compliance theory answers these questions without assuming that other inquirers will comply with their obligations or that the environment is epistemically hospitable. Absent these assumptions, we can no longer assume that a dialogue between two inquirers with opposing views will take the same sort of shape. It may become the sort of 'dialogue' where one side does not listen to the other (or worse, neither side listens to the other) and no real arguments are offered.

It is not hard to see why, because of this, full and partial compliance theory might come apart on the question of whether we should engage with those with whom we disagree. The full compliance theorist may well end up extolling the epistemic benefits of frank and open discussion—and they are especially likely to do this if they lose sight of the fact that they are relying on certain idealizations. In contrast, the partial compliance theorist may insist that, at least for many of us, these benefits are unlikely to materialize. My point is not that partial compliance theory is right and full compliance theory is wrong here (though I argue for this conclusion in Chapter 6). My point here is just that it is plausible that these two approaches will come to different answers.

2.3 Utopian vs. Realistic Theory

In §2.1, I indicated why Valentini's distinction between utopian and realistic theory might not be useful for my purposes. In this section, I want to substantiate these brief remarks. I hasten to add that I take no stand on whether her distinction is a useful way of understanding the debate in political philosophy—for all that I say here, it may well be useful.

Let me start with a distinction I drew in Chapter 1 between two 'branches' of epistemology. The first branch is centrally concerned with things like the nature of knowledge and justification. Let us call this branch the 'theory of epistemic ideals' (because it is interested in epistemic ideals). The second branch is inquiry epistemology, which I have already discussed in §2.2. The theory of epistemic ideals is a suitable candidate for being the epistemological analogue of Valentini's utopian theory. For Valentini, the utopian theorist is interested in justice rather than the question of how to make the world (more) just. Similarly, the theorist of epistemic ideals is interested in knowledge and other epistemic ideals rather than the question of how to get knowledge, or the question of how to ensure it is properly distributed within a community.

Inquiry epistemology, on the other hand, is a viable candidate for being the epistemological analogue of realistic theory. Unlike the utopian theorist, the realistic theorist is interested in the question of how to make the world (more) just. Similarly, the inquiry epistemologist is interested in the question of how to get knowledge and how to ensure it is properly distributed.

You might quibble about how good the analogy between utopian theory and the theory of epistemic ideals is. In particular, you might wonder whether the theorist of epistemic ideals is more concerned with feasibility and other real-world constraints than Valentini's utopian theorist is. We are told that the utopian theorist is not particularly interested in feasibility. They may consider, for example, the limitations of human altruism. But they will not reject a theory of justice purely because justice, as the theory construes it, is very hard to attain.

In contrast, you might think, the theorist of epistemic ideals is more concerned with feasibility. For example, it is a standing presumption in debates in the theory of knowledge that a theory of knowledge must avoid certain sceptical conclusions. Most obviously, it needs to avoid the conclusion that we have little knowledge about the external world.[3] It may be that the theorist of epistemic ideals is more concerned with feasibility than Valentini's utopian theorist is. Still, they do seem to share a common concern with the ideals themselves, rather than with what you must do to realize them. (I say a little more about the theory of epistemic ideals in Chapter 8, where I discuss an approach to it that has recently been defended by Robert Pasnau in his 2017 book *After Certainty*.)

What about inquiry epistemology as an analogue of realistic theory? The problem here is that, even if inquiry epistemology is analogous to realistic theory, the analogy does little to advance our understanding of the difference between ideal and non-ideal theory in epistemology. Realistic theory as Valentini

---

[3] It is difficult to cite any one source for this claim precisely because it is so central to the contemporary theory of knowledge. The best place to look is in the literature on responses to scepticism, where you find versions of this presumption stated explicitly. See, for example, Pritchard (2002).

understands it is a broad church. Really, we need to distinguish between more and less idealized versions of it. Similarly, inquiry epistemology is a broad church, and we need to distinguish between more and less idealized versions of it. You might pursue inquiry epistemology yet make certain assumptions and idealizations about inquiry and inquirers.

This is where Mills' discussion of ideal and non-ideal theory is useful. We saw in §2.1 that Mills has highlighted certain kinds of idealizations that he thinks are characteristic of ideal theory. We can distinguish between different realistic theories (in Valentini's sense) in terms of which of these idealizations they make and which they eschew. But, if we do that, then we are really using Mills to make sense of the difference between ideal and non-ideal theory. In the next section, I will show how we can use epistemological analogues of these idealizations to make sense of the difference between ideal and non-ideal epistemology.

2.4 Mills and Ideal Epistemology

We saw that Mills highlights four kinds of idealizations as characteristic of ideal theory. I suggest that we can understand ideal theory in epistemology as an approach to epistemological issues and questions that makes analogous idealizations. The idealizations highlighted by Mills were:

(1) Idealizations about agents and the interactions between them.
(2) Idealizations about human psychology.
(3) Idealizations about social institutions.
(4) Idealizations about our environment.

A case can be made that all these forms of idealization are prevalent in both traditional and, more importantly, social epistemology. (This is particularly so with the fourth form, which is already of clear relevance to epistemology.) But some caveats first. My claim is that idealizations like this are common in epistemology, not that all epistemologists make them. Further, my claim is not that all work in epistemology that makes these idealizations is problematic. I do not treat 'ideal epistemology' as a pejorative (see §2.5). My present aim is just to provide evidence that these sorts of idealizations are made, not to argue that it is problematic that they are made. Finally, this section assumes some familiarity with central ideas and trends in twentieth- and twenty-first-century epistemology. The reader who is not familiar with the ideas and trends I mention should consult the references provided.

Let us take each form of idealization in turn. First, epistemic agents are socially situated—they have social identities, occupy social roles, engage in extensive social interaction, and so on. Where traditional epistemology ignored our social situatedness almost entirely, social epistemology foregrounds certain aspects of our social situatedness, such as the extent to which we rely on others for the information we need. But, as critics of social epistemology have pointed out, many social epistemologists work with very 'thin' models of what socially situated epistemic agents are like (Anderson 2017; Grasswick 2018; Kusch 2010). These models are 'thin' in the sense that they ignore all sorts of aspects of our social situatedness.

Feminist epistemologists, in particular, have urged the importance of 'thicker' models of socially situated epistemic agents. For feminist standpoint theorists, for example, our social identities (our class, our race, our gender) are epistemologically relevant because different social positions afford access to different bodies of evidence (Collins 1986; Harding 1995; Hartsock 1983). Feminist epistemologists have more generally emphasized the importance of power relations to epistemology, and in particular to the epistemology of testimony (Berenstain 2016; Dotson 2011, 2014; Fricker 2007; Medina 2012; Rolin 2002). One way in which power differentials can manifest themselves is in the different levels of credibility afforded to different social identity groups. When members of a group systematically receive less credibility than they are due, we can talk of a distinctively epistemic form of injustice (Fricker 2007).

Second, traditional epistemology paid little attention to cognitive and social psychology. While an increasing number of epistemologists draw on psychology in their work, a common theme in this work is that the available psychological evidence makes trouble for some prominent views in epistemology. Here are three examples of what I have in mind (there are many more).

- Mark Alfano's work on the situational factors that influence our conduct as inquirers calls into question some of the fundamental assumptions of virtue epistemology, in particular that we have underlying intellectual character traits (Alfano 2012, 2014; Alfano and Fairweather 2017).
- Quassim Cassam's work on self-knowledge charges standard accounts of self-knowledge with ignoring our susceptibility to certain forms of cognitive bias (Cassam 2014).
- Hilary Kornblith's work on reflection charges many epistemologists with inflating the value of reflection and ignoring work in psychology that reveals its limitations (Kornblith 2012, 2019).

The fact that those epistemologists who take psychology seriously often use it as a vantage point from which to criticize standard views in epistemology is good evidence that at least some epistemologists have not taken psychology as seriously as they should have. Insofar as the targets of these critiques work in social epistemology (e.g. responsibilist virtue epistemology), this point applies to some strands in social epistemology too.

Third, traditional epistemology, and some work in social epistemology, ignores social institutions almost entirely. But those social epistemologists who do discuss social institutions often rely on idealized versions of them or focus on constructing idealized versions of our actual social institutions. For example, in 'political epistemology' it is common to claim—or at least take the idea seriously—that democracy and democratic institutions maximize epistemic goods such as knowledge and truth (Anderson 2006; Cohen 1986; Landemore 2013). This may be true of ideal versions of these institutions; it is far less likely that it is true of actual examples of them (Hannon 2020).

For another example, we can look at Philip Kitcher's influential work in philosophy of science (Kitcher 2001, 2011). Kitcher seeks to articulate an idealized version of science and scientific institutions, the idea being that we can then measure the adequacy of actual scientific institutions in terms of 'fit' with this ideal. On this picture, we can justify the structure of actual scientific institutions as approximations to this ideal—and criticize the structure as far as they are bad approximations. But, as some have argued, Kitcher seems to run into exactly the sorts of problems the ideal theorist in ethics and politics runs into. His project doesn't provide any basis for understanding the shape and structure of actual scientific institutions, and an idealized version of science is not a good 'yardstick' for assessing actual scientific institutions (Keren 2013).

Fourth, the epistemology of testimony is a core area in social epistemology. A central theme in the epistemology of testimony is that we get much of our knowledge from others. While we sometimes get testimonial knowledge from people we know or have a personal relationship with, we often get it from information sources (papers, websites, TV, etc.) and our broader 'information environment' (Goldberg 2010). But the many ways in which our information environment is less than ideal has, until recently, received surprisingly little attention. It has not escaped the attention of epistemologists of testimony that people sometimes lie. But the possibility of more organized forms of deception, or that deception might serve certain political agendas and purposes, has barely registered as a phenomenon worth investigating. Those who are familiar with contemporary social epistemology will know that, in the past few years, this has changed. But those driving this change are clearly doing non-ideal theory rather than ideal theory (Anderson 2011; Dentith 2016; Levy 2019; Lynch 2016; Nguyen 2020; Rini 2017).

Rather than provide further examples, let me stop here. If I am right, then a distinction between ideal and non-ideal epistemology that is based on these idealizations cuts across the familiar distinction between traditional and social epistemology. One sort of approach to social epistemology is ideal because it makes the sorts of idealizations highlighted by Mills. Another is non-ideal because it eschews these idealizations. (It is, of course, possible to make some idealizations but eschew others. Recall my remarks about ideal and non-ideal epistemology being best understood as tendencies.)

Let me now take up some loose ends. First, in §2.1, I suggested that Mills' distinction between ideal and non-ideal theory subsumes Rawls' distinction between full and partial compliance theory. Much the same seems to go for epistemological analogues of these distinctions. Full compliance theory in epistemology makes some of the idealizations I have highlighted in this section, that is, that epistemic agents will comply with their epistemic obligations and that the environment is epistemically hospitable. Partial compliance theory avoids these idealizations.

Second, Mills makes two criticisms of ideal theory. The first is that the ideal theorist will inevitably lack the tools to explain and understand certain things because they are not part of their idealized models. The epistemological analogue of this criticism is that, if we build idealizations about epistemic agents, their psychology, social institutions, and the information environment into our epistemological theorizing, we will inevitably lack the tools to understand and explain the epistemological consequences of the ways in which our world departs from these idealizations. Mills articulates this point nicely in this passage from 'White Ignorance':

> [T]he *potential* of [social epistemology] for transforming mainstream epistemology is far from being fully realized. And at least one major reason for this failure is that the conceptions of society in the literature too often presuppose a degree of consent and inclusion that does not exist outside the imagination of mainstream scholars... The concepts of domination, hegemony, ideology, mystification, exploitation and so on that are part of the lingua franca of radicals find little or no place here. In particular, the analysis of the implications for social cognition of the legacy of white supremacy has barely been initiated.
>
> (Mills 2007, p. 15)

Mills was writing in 2007 and the situation has improved. Even today, though, much social epistemological theorizing presupposes an idealized picture of what society is like of the sort that Mills is criticizing.

Mills' second criticism of ideal theory was that it was not only incomplete but also inadequate as a normative theory because it runs the risk of worsening our situation by taking us further away from the ends we profess to care about. The epistemological analogue of this criticism is that, if we build the sorts of idealizations that are typical of ideal epistemology into our epistemological theorizing, we run the risk of building an inadequate epistemology. As I discussed in Chapter 1, the target of this criticism is not so much the epistemologist who is interested in knowledge or justification (the theorist of epistemic ideals) but rather the inquiry epistemologist who is interested in how we should conduct our inquiries yet relies on an idealized picture of what inquirers are like. In building various idealizations into inquiry epistemology, we run the risk of making suggestions for improving inquiry that will not serve our epistemic ends because they are predicated on false assumptions about the situation in which we find ourselves.

To put this in more concrete terms, here is an example. What I say here anticipates the argument of Chapter 5, so if you do not find this convincing, I hope to change your mind later. One idealization we might make in characterizing epistemic agents is to ignore their tendency to engage in motivated reasoning. Put roughly, motivated reasoning is reasoning that serves goals, like the goal of maintaining a positive self-image or promoting the interests of your social group, that have little to do with the accuracy of what you believe (Ellis 2022). This idealization becomes problematic if we consider the merits of a norm of inquiry that tells inquirers to 'do their own research'. If we assume that inquirers are (for the most part) trying to form an accurate view about the matter at hand, then a norm like this might look sensible. On the other hand, once we recognize that inquirers are often not doing this, it becomes harder to make the case that doing your own research will be an effective way of uncovering truths and gaining knowledge, whether about yourself or anything else. It may well be that, when you do your own research into a political issue, all you do is gather more evidence that supports a stance on the issue that serves the interests of your group. Indeed, you might think that doing your own research will often lead to you doing quite badly from the epistemic point of view—worse than you would have done if you had been more willing to accept help from others.

2.5 Objections

In this chapter, I have identified three ways of distinguishing between ideal and non-ideal epistemology and suggested that the most useful way of drawing the distinction is, following Mills, in terms of certain idealizations that are typical of ideal epistemology and eschewed by non-ideal epistemology. Most of the chapters that follow make use of this way of drawing distinction, though in Chapter 6 I also make use of Rawls' distinction between full and partial compliance theory. I will finish this chapter by considering some objections to non-ideal epistemology and my overall project. In the process, I will clarify some aspects of non-ideal epistemology and further prepare the ground for the chapters that follow.

Objection 1: The non-ideal epistemologist is interested in what is the case—what inquirers are really like, how they really think, what our actual social institutions are like, and so on. But they do not just want to develop a detailed characterization of our situation as inquirers. Their aims are ameliorative. The non-ideal epistemologist wants to improve inquiry, not just describe it. But then they face the problem that you cannot derive an 'ought' from an 'is'. You cannot derive normative claims about what inquiry should be like from descriptive claims about what it is really like. The non-ideal epistemologist either needs to stick to making descriptive claims or to engage in some normative theory. But—so the objection goes—if they engage in normative theory then they are no longer doing non-ideal epistemology.

Response: This objection is reminiscent of the standard objection to Quinean naturalized epistemology, which is that you can't derive a normative epistemology from psychology (Kim 1988). While it has been elaborated on by other naturalized epistemologists (e.g. Kornblith 1993), Quine's own response to this objection strikes me as fundamentally right. Quine held that naturalized epistemology was akin to engineering in that it asks what humans need to do to satisfy the goal of having true beliefs (Quine 1986). The norms that the naturalized epistemologist recommends are 'normative' in the way that all instrumental norms are normative. They tell us what to do to achieve a desired goal. The non-ideal epistemologist can respond to the first objection in much the same way. The non-ideal epistemologist is engaged in a form of engineering. They want to design norms of inquiry and 'epistemic environments' that will be effective in producing knowledge and other epistemic goods. Their norms are normative in the way that the naturalized epistemologist's norms are normative. They are instrumental norms, which tell us what to do to achieve a desired goal.

Objection 2: You might respond that what I have just offered is an instrumentalist account of the norms proposed by non-ideal epistemology. But many have objected that instrumentalist accounts of epistemic normativity are problematic because whether it serves our purposes to believe something is irrelevant to whether we have any justification for believing it (e.g. Berker 2013; Kelly 2003).

Response: Whether or not this is a good objection to instrumentalist accounts of norms governing belief, the non-ideal epistemologist need not worry about it. The non-ideal epistemologist is interested in norms of inquiry and, while this is controversial, I do not view the norm of belief as a norm of inquiry. By 'norm of inquiry', I mean a norm that tells actual inquirers what they should do in their inquiries—whether they should gather more evidence, whether they should reach a conclusion or keep on inquiring, and so on. Norms like this are *prescriptive*: we are meant to do what they enjoin us to do. In contrast, as Mona Simion, Chris Kelp, and Harmen Ghijsen (2016) have argued, the norm of belief is *evaluative* rather than prescriptive. It is a 'norm' that evaluates beliefs qua beliefs (tells us what it is for a belief to be a good belief) rather than a norm that tells us how to go about forming beliefs. (Compare: the 'norm' that a good knife should be sharp tells us what it is for a knife to be a good knife, not that we should go and sharpen our knives.)

My suggestion, then, is that while instrumentalism about evaluative epistemic norms may be untenable for reasons given by Kelly and others, I see no reason for thinking instrumentalism about prescriptive epistemic norms is untenable. Indeed, it seems like an eminently sensible view. Surely, whether you should gather more evidence, or whether you should gather a different kind of evidence, depends at least in part on whether doing so will serve your epistemic goals. Do you need more (or a different kind of) evidence to have knowledge, for example?

Objection 3: A standard objection to non-ideal theory in political philosophy is that, if the non-ideal theorist wants to say anything about what we should do, they need to rely on ideal theory to tell them what they are trying to achieve. For example, if the non-ideal theorist wants to say something about how we should tackle injustice in the here-and-now they need to have a conception of what a just world would look like. But only ideal theory can provide this conception. As Rawls puts it, the ideal part of his theory 'presents a conception of a just society that we are to achieve if we can' (Rawls 1971, p. 216). A parallel point can be made about ideal and non-ideal epistemology. We need to know what our epistemic goals or ideals are before we can go about asking how to achieve them and only ideal epistemology can tell us what our epistemic goals are.

Response: I want to make two points in response. First, even if this is right, it hardly shows that non-ideal epistemology is unimportant or unnecessary. You can be interested in what I call the theory of epistemic ideals *and* the normative question of how we, as inquirers, should go about our task in the situations in which we find ourselves (i.e. inquiry epistemology). The theory of epistemic ideals supplies a conception of what we are aiming at—whether we are aiming at truth or knowledge, what knowledge is, and so on. The inquiry epistemologist then asks how we, both as individual inquirers and as a community, might go about achieving these aims. So even if inquiry epistemology must presuppose some results from the theory of epistemic ideals (i.e. what it is we are aiming at in our inquiries) this hardly shows that inquiry epistemology in general, or a non-ideal version of it, is redundant.

Second, we can push back against the idea that you cannot go about tackling injustice without having some idea of what a just world would look like. You need to know what you are aiming at in some minimal sense. You need to know that you are aiming to make the world (more) just, for example. But this hardly requires having a theory of justice. You do not need a theory of justice to know that certain things, such as racial discrimination, are unjust. You also do not need a theory of justice to start thinking about how to prevent racial discrimination. As Elizabeth Anderson puts it:

> Knowledge of the better does not require knowledge of the best. Figuring out how to address a just claim on our conduct now does not require knowing what system of principles of conduct would settle all possible claims on our conduct in all possible worlds, or in the best of all possible worlds.   (Anderson 2010, p. 3)

A parallel point applies in epistemology. You need to have some conception of what knowledge is to aim at the production of it. For example, you might need to know that knowledge is factive (you cannot know something that is false). But you do not need to have a particularly firm conception, certainly not the sort of conception that is at issue in the theory of knowledge.

To make this point more concrete, consider an example. In some countries a sizeable proportion of the population are sceptical about the safety of vaccines. These people lack some knowledge (i.e. that vaccines are safe) that it would be good, both for them and for everyone else, to have. The non-ideal epistemologist will ask what we can do to improve the epistemic situation of these vaccine sceptics. Addressing this question will require finding out what drives vaccine scepticism (why do they think this?) and looking at empirical research on which sorts of interventions are most effective in changing the minds of vaccine sceptics. Throughout this process, the non-ideal epistemologist needs some idea of what they are trying to achieve. At the very least, they want to stop some people believing false things about vaccines. If they are more ambitious, they want people to come to recognize—to know—that most vaccines are safe. But for this purpose, they do not need anything approaching a theory of truth or knowledge. All they need is the sort of understanding of truth or knowledge possessed by any competent user of the word 'true' or 'knows'.

Objection 4: You might object that, while what I am calling non-ideal epistemology is important, it is already being done. Educational theorists and policy-makers want to figure out how to provide young people with an excellent education. While there is more to an excellent education than imparting knowledge, their aim is at least partly to promote certain epistemic goods, and their methods are clearly closer to those of non-ideal than ideal epistemology. Psychologists do not (just) investigate cognitive biases because they want to document the myriad ways in which actual human cognition falls short of some imagined cognitive ideal. Their work has an ameliorative aspect. You cannot formulate strategies for minimizing the influence of biases when you do not understand the biases themselves or why we are so prone to them. Because some biases lead people to miss out on epistemic goods (to form false beliefs, to misunderstand things), the ameliorative aspects of psychological work on biases will also look a lot like non-ideal epistemology.

Response: Let me distinguish two versions of this objection. On the first version, the point is that non-ideal epistemology is already being done so it is not incumbent on epistemologists to do it. There is a clear division of labour: epistemologists do what I am calling ideal epistemology (or the theory of epistemic ideals), and others (educational theorists, psychologists) do what I am calling non-ideal epistemology. My response to this version of the objection is just that epistemologists can *also* do—indeed already do—non-ideal epistemology. It might help to reiterate that I am not calling for epistemologists to abandon ideal epistemology and *only* do non-ideal epistemology. I am calling for more recognition of the shortcomings and limitations of ideal epistemology. If they want to properly address some of the issues they discuss, epistemologists need to do some non-ideal theory.

On the second version, the point is that epistemologists are not well placed to do non-ideal epistemology. We are too wedded to abstractions and idealized ways of thinking. We lack the training to do non-ideal theory well. My response to this version of the objection is curt. There are examples of good non-ideal epistemology done by epistemologists (see §2.4). (I hope this book will also be an example of good non-ideal theory done by an epistemologist!) However, for the avoidance of doubt, I certainly do not claim that only epistemologists can do non-ideal epistemology, or that epistemologists do it better than others.

Objection 5: Ideal epistemology is extremely useful. To take just one example, Bayesian epistemology can certainly be described as a form of ideal epistemology (Carr forthcoming). But it has been influential in a range of disciplines including medicine, computing, AI, and finance (Lin 2022). This is just one example of the more general point that abstraction and idealization are not necessarily in tension with practical applicability. Even the most abstract and esoteric disciplines can have practical uses. (Moreover, the less abstract disciplines are often of less practical use than their practitioners seem to think.) You might object that my critique of ideal epistemology ignores this crucial point.

Response: I completely agree with all of this. There need be no tension between idealization and practical applicability and Bayesian epistemology may well be a case in point. There are, however, two important things to recognize here. First, as Mills emphasizes, abstraction and idealization always serve to obscure certain phenomena. If I abstract away from friction and consider a frictionless plane when constructing my mechanical theory, then my mechanical theory will not be able to explain why things get hot when they rub against each other. Whether this is a problem depends on several things, most importantly whether I want my theory to explain why things get hot when they rub against each other. (We clearly want to explain this, but we might not want my theory to explain it.)

More generally, whether there is anything wrong with abstraction or idealization depends on whether we are unable to explain what we need to explain because of our idealizations. Mills' critique of ideal theory in ethics and politics is precisely that it obscures phenomena we want to explain and that you would expect ethics and political philosophy to explain, such as the persistence of racial discrimination. The non-ideal epistemologist's complaint about ideal epistemology is similar: it obscures phenomena that need explaining and we would expect epistemology to explain. But this does not mean it is incumbent on every epistemologist to explain these phenomena, or that only non-ideal epistemology can be practically relevant.

Second, as Mills also emphasizes, in theorizing about a system you try to only abstract or idealize away from aspects of the target system that do not matter for your purposes. This is a characteristic of what philosophers of science call *minimalist idealization*, which is 'the practice of constructing and studying theoretical models that include only the core causal factors which give rise to a phenomenon' (Weisberg 2007, p. 642). This yields a 'minimalist model' of the phenomenon, which 'contains only those factors that *make a difference* to the occurrence and essential character of the phenomenon in question' (Weisberg 2007, p. 642).

We can helpfully view Mills' critique of ideal theory in these terms. For Mills, the ideal theorist is engaged in something like minimalist idealization but ends up constructing models that ignore factors that do make a difference to the character of the phenomena they want to study. The point, then, is that the ideal theorist, whether in ethics or epistemology, abstracts away from things that make a difference to the phenomena they are interested in. The non-ideal epistemologist's complaint about ideal epistemology is that it abstracts away from things that make a difference when we consider how we should structure our inquiries. A norm of inquiry might look like it will serve our epistemic goals in the abstract, but once we recognize that we have abstracted away from certain aspects of our actual situation it may become clear that, in practice, it will only frustrate our epistemic goals.

Objection 6: You might object that, as far as non-ideal epistemology styles itself as a critique of 'mainstream' analytic epistemology, it misses its target. Most analytic epistemologists are engaged in things like the theory of knowledge or justification. But I have explicitly said that I am not targeting things like the theory of knowledge or justification.

Response: Clearly, many analytic epistemologists are primarily concerned with the theory of knowledge or justification. But many analytic epistemologists also engage in inquiry epistemology and some of them do so in a recognizably ideal way.[4] This point is best made through some examples. The theory of justification is certainly *relevant* to the question of how to respond to peer disagreement (a central question in social epistemology). But you cannot properly answer the question of how to respond to peer disagreement just by laying down the conditions under which a belief, formed in response to peer disagreement, would be justified or count as knowledge. It is one thing to say that a belief is or would be justified and quite another to say what someone should believe given the evidence at their disposal. (This goes back to my earlier point that the norm of belief is evaluative rather than prescriptive. If this is right, it is a mistake to expect a theory of justification to say what you should or should not believe.)

The point is even clearer when you consider questions about the gathering of evidence ('how should you gather evidence?', 'which information sources should you consult?'), expert deference ('which experts should you trust?') or the myriad

---

[4] I therefore don't agree with Ballantyne's (2019) contention that the vast majority of contemporary epistemologists aren't interested in inquiry epistemology (or, as he calls it, regulative epistemology). Much the same goes for Pasnau's (2013) contention that epistemologists are too enamoured with descriptive questions ('what is knowledge?' and the like). Both contentions may have been true about epistemology in the last few decades of the twentieth century, but I do not think they capture the sheer range of different projects in contemporary social epistemology (see the examples mentioned in §2.4).

norms governing communal inquiries and the interpersonal relationships within them ('who should I engage with?', 'which criticisms are worth taking seriously?'). These are normative questions about how to conduct our inquiries, not questions about what justification is. A theory of justification may inform our answers to them, but it cannot itself supply those answers. For that, we need inquiry epistemology. In the chapters that follow, I argue that, if we want to answer questions like these properly, we need a non-ideal form of inquiry epistemology.

Objection 7: You might worry that my objection to ideal epistemology makes an egregious mistake. Stripping away the complexities, it might seem like I am arguing that we sometimes do a bad job of following the norms that the ideal epistemologist says we are under. This may be true, but it hardly shows that we are not beholden to those norms. That you can try to follow a norm yet do a bad job of doing so does not show that you are not beholden to the norm. It just shows that you can fail to do what it asks of you.

Response: Let me make two points in response. First, the non-ideal epistemologist's argument is not that we sometimes fail to do what the ideal epistemologist says we should do. The argument is more like this. If we want to achieve our epistemic goals (e.g. knowledge), then we need to consider whether trying to follow this or that candidate norm is likely to be conducive to achieving those goals. If it turns out that it is not likely to be conducive to those goals, then we should reject it as a bad norm. The claim, then, is that some of the norms proposed by the ideal epistemologist are like this. If we try to follow them, we are likely to frustrate our epistemic goals. For example, imagine that trying to follow a norm enjoining us to 'do our own research' (to be intellectually autonomous) is likely to lead to bad epistemic consequences—to a worsening not an improvement of our epistemic situation. This would be an argument against this candidate norm of inquiry: it does not serve the epistemic goals that we have.

Second, the non-ideal epistemologist does not just offer a critique of ideal epistemology. The non-ideal epistemologist offers an *alternative* to it. They want to propose *different* norms—norms that are more likely to achieve our epistemic goals. My argument does not rely on the implausible claim that, because ideal epistemology traffics in norms that are hard to follow, it should be rejected. Rather, it relies on the plausible claim that, at least with respect to some epistemological issues, the norms proposed by the non-ideal epistemologist are more likely to serve our epistemic goals than the norms proposed by the ideal epistemologist.

Objection 8: My objection to ideal epistemology might not make an egregious mistake. However, you might worry that my objection to ideal epistemology is only cogent insofar as my target is guilty of an egregious mistake. Some ideal epistemologists might go too quickly from conditional claims about how we should inquire *if* certain idealized assumptions are met to unconditional claims about how we should inquire. This certainly is a mistake. From the fact that we would be under a norm or obligation if certain conditions were met it does not follow that we are under the norm because the conditions may not be met. But the fact (if it is a fact) that some 'ideal epistemologists' make this mistake is not an argument against ideal epistemology so much as an argument against *bad* ideal epistemology.

Response: Stated like this, the mistake does look egregious. However, there is a reason someone might make such an egregious mistake (a reason that does not require that they be a bad philosopher!). One reason you might go from the claim that we *would* be under a norm or obligation if certain conditions are met to the claim that we *are* under that norm is if you think that the conditions are met, at least in their essential respects. For example, you might recognize that actual discourse is a little different from a philosopher's idealized version of it while also thinking that actual discourse is *close enough* to the idealized version that you can still derive normative conclusions from the idealized version. In the chapters that follow (especially Chapters 5 and 6) I argue that some ideal epistemologists make exactly this sort of mistake. For example, someone who is attracted to a Millian picture of the value of debate may recognize that actual debate is a little different from how Mill imagines it. But they may still think that it is close enough to what Mill imagines that the claims he makes about why we should engage in debate stand up.

Let me make another point here. It is tempting to argue that ideal epistemology is in a privileged position vis-à-vis non-ideal epistemology. Only ideal epistemology can deliver a picture of what an epistemically ideal agent (or epistemic community) would look like, and you might think that we need such a picture to know what we are working towards (see Objection 3). I think this argument can be resisted: you do not need a particularly complete picture of what you are working towards to work towards it (see my response to Objection 3). But if we admit (as the objection under discussion does) that the ideal epistemologist is only in the business of making conditional normative claims (claims about what we should do if certain conditions are met), this argument does not even get off the ground. As it turns out, ideal epistemology is not in the business of telling actual inquirers what they should do.

Objection 9: In a discussion of ideal and non-ideal theory in the philosophy of language, Herman Cappelen and Josh Dever argue against importing the distinction between ideal and non-ideal theory into the philosophy of language (Cappelen and Dever 2021). Their main contention is that the various versions of the ideal/non-ideal distinction boil down to the platitude that, in any field, there are topics that are under-explored. Because you do not need to invoke a distinction between ideal and non-ideal theory to make this point, they regard the distinction as superfluous, at least in the philosophy of language. While their paper is about philosophy of language, not epistemology, you might think that a parallel case can be made against importing the distinction into epistemology.

Does the distinction mark anything more than the platitude that some topics are under-explored in epistemology?

Response: The three versions of the ideal and non-ideal epistemology distinction I have considered in this chapter do not boil down to the platitude that there are some underexplored issues in epistemology. That said, it is important to underscore why they do not boil down to this platitude. Mills is, again, helpful here. Let us start with Mills' point that ideal theory tends to obscure certain phenomena. The claim here is not just that ideal theory does not address every issue or question of interest. The claim is that, from the vantage point of ideal theory, certain issues or questions are obscured precisely because we have idealized away from the features of the world that might lead us to consider these issues or pose these questions. The point is that these issues are under-explored not by accident but, as it were, by design.

Mills' second point was that, insofar as ideal theory addresses normative questions (what we should do, how we should inquire), it is liable to get things wrong. The claim here is that, at least sometimes, ideal and non-ideal theory are addressing the same (normative) questions but give different answers. Only one of these answers can be right and, for Mills, the ideal theorist often has the wrong answers. You can, of course, argue that ideal theory is a better guide to action (or inquiry) than Mills supposes. But his argument does not boil down to the platitude that there are some under-explored issues in ethics and politics (or epistemology).

Let me recap. I started this chapter by drawing a contrast between epistemology and ethics/political philosophy. While it is often said of both fields that they are too abstract and idealized, there is a heated debate in ethics and political philosophy about ideal theory, while there is no parallel debate in epistemology. One of my aims in this chapter has been to demonstrate that the existing debate about social epistemology is not a surrogate for debate about ideal theory in epistemology. You can do a lot of ideal theory while doing social epistemology.

My other aim has been to explain how the distinction between ideal and non-ideal epistemology might be drawn. I have identified three ways of drawing it:

(1) Ideal epistemology as full compliance theory and non-ideal epistemology as partial compliance theory.
(2) Ideal epistemology as utopian theory (the theory of epistemic ideals) and non-ideal epistemology as realistic theory (inquiry epistemology).
(3) Ideal epistemology as characterized by certain idealizations and non-ideal epistemology as characterized by the avoidance of these idealizations.

I have argued that the third way of drawing the distinction is the most useful and I will appeal to it in the following chapters. But I will occasionally make use of the first way of drawing the distinction, particularly in Chapter 6.

The rest of this book takes up various issues and questions in social epistemology. In each case, my aim is to argue for a non-ideal approach to the issue or
question. I start with the question of how to identify experts. While many of us are
experts in some fields, nobody is an expert in all fields. We all need a way of
figuring out who to listen to and get information from in fields about which we are
ignorant. I will argue that a prominent approach to this issue, that taken by Alvin
Goldman, is problematic because it involves precisely the sorts of idealizations
highlighted by Mills. This will lead me to a non-ideal alternative to Goldman's
approach, and to a defence of what I called in Chapter 1 the 'institutional face' of
non-ideal epistemology.

3

Anderson and Goldman on
Identifying Experts

We know some things, but there is a lot we do not know. Sometimes, the fact that we do not know something does not matter because we do not need to know it. I do not know much about crystallography, but this does not matter because I do not need to know anything about crystallography. But, other times, the fact that we do not know something does matter. Imagine I want to improve my diet. If I do not know what a good diet looks like, I am going to find this difficult. What I need is an expert—someone who can tell me what a good diet looks like. But now I face a new problem. Lots of people claim to be experts about what a good diet is, and they often say different, indeed contradictory, things. What can I, as someone who knows little about this topic in the first place, do in this situation?

In this chapter, I compare two approaches to this problem, which I will call the 'problem of identifying experts'. The first is developed by Alvin Goldman in his 2001 paper 'Experts: Which Ones Should You Trust?'; the second is developed by Elizabeth Anderson in her 2011 paper 'Democracy, Public Policy, and Lay Assessment of Scientific Testimony'. I argue that Anderson's approach to the problem is preferable to Goldman's because it is less ideal. Indeed, it is a good example of non-ideal epistemology in action.

This chapter tackles the problem of identifying experts in an indirect fashion, by contrasting two approaches to the problem. I adopt this indirect approach because I have two aims. The first is to argue for a non-ideal approach to the problem of identifying experts. The second is to illustrate the contrast between ideal and non-ideal epistemology via a detailed case study.

Here is the plan. I start by outlining Goldman's approach to the problem of identifying experts (§3.1) and explaining why it is an example of ideal epistemology (§3.2). I then consider Anderson's approach (§3.3). While Anderson starts as Goldman does by identifying criteria you might expect to be useful for identifying experts, she then argues that these criteria will be less useful than you might expect in view of the sorts of facts about human cognition and our information environment that Goldman idealizes away from.

With this in hand, I argue that Anderson's approach to the problem of identifying experts is preferable to Goldman's (§3.4). Where Goldman sees the problem as requiring a solution at the level of individuals—how can I, as an individual inquirer, go about identifying experts who have the information

Non-Ideal Epistemology. Robin McKenna, Oxford University Press. © Robin McKenna 2023.
DOI: 10.1093/oso/9780192888822.003.0003

I need?—Anderson sees the problem as requiring a solution at the societal and institutional level. Rather than asking how an individual inquirer can go about identifying experts, Anderson asks how we can construct better systems of knowledge production and dissemination. Her thought is that, because people are always going to exhibit various biases in their decisions about who to get their information from, we need to find ways of enabling them to make the right decisions. This will mean constructing an information environment in which it is easier to make the right decisions, and harder to make the wrong ones. In effect, it means doing non-ideal institutional epistemology.

I finish by highlighting two key tasks that need to be carried out to fully defend Anderson's non-ideal approach to institutional epistemology (§3.5). The first is to fill in some of the messy empirical details. Which steps might we take to construct a better information environment? What evidence do we have that these steps would work? The second task is to deal with a fundamental objection to Anderson's approach. The objection is that what Anderson proposes is (arguably) a form of paternalism and paternalism is, in general, deeply problematic because it infringes on our autonomy. As I will suggest, this is a general objection to institutional epistemology, and it requires detailed treatment. I take up both these tasks in the two chapters that follow this one.

3.1 Goldman on Identifying Experts

In this section, I outline the approach Goldman takes to the problem of identifying experts. To simplify matters, Goldman asks us to consider the following situation. Layperson L wants information about a topic T that they know little about. For instance, they want to know about global warming—are global temperatures rising, and if so, is human activity the cause? They are aware of two (putative)[1] experts about T, A and B, who assert many contradictory things about T. For example, A says that global temperatures are rising, and that human activity is the main cause. B says that global temperatures are not rising and, even if they were, human activity would not be the cause of it. Goldman's question is: how should L decide which of A and B is more credible?

Before getting to his answer, it will be helpful to discuss a few more features of the way Goldman frames the problem. First, Goldman takes it that it is, at least sometimes, permissible to rely on expert testimony. Indeed, he takes it that it is, at least sometimes, the only way we are going to get the required information. He rejects the view, which might be defended on the grounds of the value of

---

[1] Goldman's question is how to decide if someone is an expert in the first place, so we cannot build it into the description of the problem that A and B are experts. I sometimes omit the 'putative' for reasons of readability, but the reader should always read 'expert' as 'putative expert'.

intellectual autonomy, that we should forswear all forms of epistemic dependence on others. (For more on intellectual autonomy, see Chapter 5.)

Second, Goldman is suspicious of the view that we may blindly rely on expert testimony. On this sort of 'blind deference' view, we would be entitled to rely on a piece of expert testimony even absent any positive reasons for doing so. In the epistemology of testimony, it is hotly debated whether we need positive reasons for relying on testimony. One side—the 'blind deference' side—says that we don't (Burge 1993; Coady 1992); the other side says that we do (Audi 1997; Fricker 1994). But Goldman does not need to enter into this debate because he focuses on cases where two (putative) experts assert contradictory things. Any entitlement we may have to blindly rely on testimony is defeasible and one way in which it would be defeated is if we were aware of two testifiers who we initially regarded as equally credible but say conflicting things.

Finally, Goldman rejects the sceptical view that we simply have no way of identifying experts. But he does not claim that the problem of identifying experts admits of an easy solution and he accepts that, sometimes, it may not admit of any solution. Relatedly, his aim is not to give us 'fail-safe' ways of deciding which expert to trust. Any criterion a layperson may use to decide between two conflicting experts can, and will, go wrong, and it can, and will, be misapplied. These points will be important when considering the viability of Goldman's approach.

With these clarifications in place, we can now look at Goldman's five criteria a layperson can use to choose between two competing experts. They are:

(1) Evaluate the arguments presented by the experts to support their own views and attack their rivals' views.

(2) Look for agreement from other putative experts on one side or other of the dispute.

(3) Look at appraisals by 'meta-experts' of the experts' expertise.

(4) Look for evidence of the experts' interests and biases that might lie behind the dispute.

(5) Consult evidence about the experts' track-records.

It will be helpful to briefly examine each criterion. The first criterion is useful when there is a debate between the rival experts in which they present arguments. Goldman's thought is that, while laypersons may lack the expertise needed to evaluate the claims made by experts, they can evaluate their arguments using the tools of formal and informal logic. Further, they can look for 'indirect indicators' that one argument is better than another (e.g. the inability of the opposing expert to respond to it in a coherent way). While neither method is 'fool-proof', they may provide defeasible reason to side with one expert over the other.

The second and third criteria both appeal to forms of consensus. The second is based on the idea that, if all or most of the other experts agree with one of the rival experts, then this is a reason to side with them. The third is based on the idea that, if all or most other experts view one of the rivals as an expert but not the other (or view one of the rivals as more of an expert) then this is also a reason to side with them. Goldman discusses the need to distinguish between consensus and 'group-think', but we do not need to get into this here. It is, however, important to note that, as with the first criterion, the existence of consensus (whether on the expert's claims, or on their status as an expert) only provides a defeasible reason to side with them.

The fourth criterion is based on the thought that, if the layperson has evidence that one of the rival experts is biased and that their claims reflect this bias, whereas the other rival expert isn't biased (or their claims don't reflect any bias they do have), then they have reason to trust the unbiased (or less biased) expert over the biased. It is true that the fact that someone is biased in favour of a particular outcome is not conclusive evidence that they are wrong. If a scientist dearly wants their experiment to establish a particular outcome this does not in itself invalidate their experiment if it supports the desired outcome. It all depends on whether their bias leads them to distort their research. But, again, the fact that one expert is biased while the other is not provides defeasible reason to side with the non-biased expert because it may make it more likely that they have distorted their research.

Finally, the fifth criterion is based on the thought that, if the layperson can look at the past track-records of each rival expert, and if it turns out that one has a clear track-record of being right whereas the other has a clear track-record of being wrong, then they have a reason to trust the one with the good track-record over the one with the bad. Of course, it may be that someone with a track-record of getting things right gets something wrong, or that someone with a track-record of getting things wrong gets something right. But this reason is, again, defeasible.

With these criteria in hand, Goldman turns to the question of whether we can expect them to be of any use in solving the problem they are intended to solve (i.e. the problem of identifying experts). Goldman's strategy here is to argue that there are situations in which they can be applied. Here is an—admittedly abstract—example of a situation where his first criterion can be applied.

Imagine you are watching a debate between two experts. One expert consistently stumbles, does not properly respond to objections, says things that seem incoherent, and is near-incomprehensible. The other expert speaks confidently, responds to all objections, is very coherent, and perfectly comprehensible. You can apply Goldman's first criterion here, side with the second expert over the first, and accept their testimony. While real-life situations are rarely as clear-cut as this, it is also true that we can often reliably judge who has performed best in a debate. Similar things can be said about Goldman's other criteria. So, while it is an open question how often Goldman's criteria can be applied, there is no good reason to think they could never be applied. Goldman is happy with this modest claim because, as I emphasized earlier, his aim is also modest. He is simply trying to show that, in certain situations, the problem of identifying experts admits of a solution. He does not claim that it admits of a solution in all situations.

You might wonder whether Goldman is worried about the fact that his criteria are not 'fail-safe'. That is, is he worried that laypeople might do a bad job of applying them? Goldman recognizes this is a possibility, but he does not think it is a problem. Goldman does not think it is a problem because, again, his aim is modest. He is not trying to show that his criteria can be applied in all situations, and he is not trying to show that anyone can apply them and get the right results. It is enough for him that, in certain situations, laypersons can make use of them to distinguish between genuine and pseudo experts.

3.2 Goldman and Ideal Theory

It is hopefully clear that Goldman's approach to the problem of identifying experts is, in some important sense, idealized. In this section, I will make this sense more precise. In fact, a case can be made that Goldman's approach to the problem of identifying experts engages in all four forms of idealization I said were typical of ideal epistemology.

First, Goldman adopts the 'thin' characterization of socially situated epistemic agents that is typical in much (social) epistemology. We are not told anything about the epistemic agents who face the problem of identifying experts, beyond that they lack certain information they need, and that they need to find a way of obtaining it. He therefore implicitly assumes that the problem has much the same shape irrespective of the particulars of your social situation—your social identity, your subjective experiences, your background beliefs, the relations in which you stand to other inquirers, and the like. This means that the particular reasons people might have to be distrustful of experts, such as a history of past abuse and deception, don't play any role in his analysis (Goldenberg 2021).

Second, he does not consider any empirical literature on how we go about identifying experts. In particular, he does not consider any of the literature that documents the various cognitive biases that influence our assessments of expertise (see §3.3). As we will see, these assessments are often deeply biased in that they tend to reflect both our existing beliefs about the matter in question and our social and political values. Put simply, we look for people who agree with us and call them experts, rather than looking for experts to tell us what to think in the first place.

Third, while Goldman discusses the conditions under which expert consensus carries epistemic weight, he does not consider the question of whether those conditions are met by a social institution like science as it is currently practised. It may be that, while there is an idealized version of science as a social institution which satisfies these conditions, science as it currently is does not satisfy them.

To determine the extent to which it does, we need to look at the actual causes of scientific consensus and try to identify conditions under which scientific consensus does and doesn't carry epistemic weight (Miller 2013). The point is not that, once we do this, it will turn out that Goldman is wrong to put so much store in consensus. The point is that whether he is right to put so much store in consensus is an empirical matter and it can only be settled by looking at how scientific consensus emerges, not at how we might imagine it does, or at how it would emerge in an idealized version of science.

Fourth, he does not consider the impact that our actual information environment has on our ability to successfully apply his criteria. In the case of some scientific issues (e.g. global warming) there is a huge amount of misinformation out there (again, see §3.3). As we will see in future chapters, there is a large empirical literature on misinformation, and one of the key results in this literature is that misinformation tends to 'drown out' genuine information (Lewandowsky et al. 2012). As a result, when misinformation about something like the degree of scientific consensus on an issue abounds, laypersons are less likely to have accurate views about what the degree of scientific consensus is.

So far, I have merely tried to substantiate the initial impression that Goldman's approach to the problem of identifying experts is an example of ideal epistemology. As I emphasized in Chapter 2, whether ideal epistemology is the right approach to take to a particular issue or problem depends on the issue. The fact that Goldman makes the sorts of idealizations just identified is not itself an objection to his approach. Sometimes idealizations are warranted, and indeed helpful in shining light on a problem. That said, in the rest of this chapter, I argue that Goldman's approach is problematic, and it is problematic because of its idealized nature. We can see why by contrasting his approach with another, less idealized, approach to the same problem.

3.3 Anderson on Identifying Experts

In her 'Democracy, Public Policy, and Lay Assessment of Scientific Testimony' Elizabeth Anderson addresses the same problem as Goldman. In this section, I give a detailed account of Anderson's approach, before identifying the crucial respect in which it differs from Goldman's, which is that Anderson's approach is an example of non-ideal epistemology. We can start with how Anderson's initial framing of the problem differs from Goldman's.

Goldman views the problem of identifying experts as a problem for individual inquirers. His question is how I, as an individual inquirer, can go about identifying those who can give me the information I need about the topics I know little about. Anderson, in contrast, views the problem as being as much a problem for society as for individuals. As she points out, responsible public policymaking in a technologically advanced democratic society relies on complex scientific research, but it must also be democratically legitimate. These demands can often seem to be in tension because the complexity of much scientific research is a barrier to public understanding, never mind acceptance, of its results, and democratic legitimation requires at least some level of acceptance of the science on which public policy is based. While solving the problem of how to identify scientific experts may not itself be enough to resolve this tension, it is at least necessary. Thus, Anderson views the problem in political as well as epistemological terms. This difference between Godman and Anderson becomes important later when I address the objection that Goldman and Anderson are not really interested in the same issue.

Let me now turn to outlining Anderson's approach. Anderson starts by considering a similar question to Goldman: how can laypersons decide which expert(s) to trust? One way is for laypersons to decide which expert(s) are *deserving* of trust. But what is it to be deserving of trust—to be *trustworthy*?

Anderson distinguishes between three aspects of trustworthiness. The first is expertise: to be trustworthy about some topic, you need to know a lot about the topic. The second is honesty: to be trustworthy about some topic, you also need to say what you honestly believe (you can know a lot about a topic but lie). The third is epistemic responsibility: to be trustworthy about some topic, you need to fulfil certain epistemic responsibilities. For instance, you need to respond to criticisms and questions from other inquirers about the topic.

She then identifies some questions a layperson can ask in assessing how trustworthy some expert testimony is, each of which speaks to one of the three dimensions of trustworthiness:

(1) How much of an expert is the testifier?
(2) How honest is the testifier?
(3) How epistemically responsible is the testifier?

As should be clear, these questions are similar to Goldman's criteria. Accordingly, I will focus on areas of overlap and divergence. Anderson's first question might look a little useless: the problem is to identify whether the testifier is an expert in the first place. But her thought is that you cannot simply split people into laypersons and experts. There is a hierarchy, ranging from the total novice to the world-leading expert. When faced with testimony by a (putative) expert, it is important to ask where the testifier stands on that hierarchy because the degree of weight you should give their testimony partly depends on where they stand. Anderson's understanding of the problem is therefore, at least in this respect, more nuanced than Goldman's.

Her second question includes Goldman's idea that laypersons should look for indications of bias, indications of bias being a reason to think someone is likely to be dishonest. But she also considers things like a track-record of misleading statements and of bad scientific practice (e.g. cherry-picking data). It therefore combines elements of Goldman's fourth and fifth criteria too.

Anderson's third question is couched in different language from all of Goldman's criteria, but indications of epistemic (ir)responsibility include things Goldman mentions, like how the testifier behaves in discussion (his first criterion), and the track-record of the testifier (the fifth criterion). For example, do they have a history of making inaccurate claims? Do they properly respond to criticism?

Anderson now moves on to consider the question Goldman considered. how easy would an ordinary layperson find it to assess the level of expertise of putative experts using these questions? It is at this point that Anderson's and Goldman's approaches to the problem of identifying experts diverge. Goldman considers idealized scenarios where the problem is tractable (scenarios where an individual can apply one or more of the criteria). Anderson considers whether, with respect to particular issues, most of us are likely to be able to use these criteria/questions to identify the 'genuine' experts and distinguish them from the 'pseudo' experts. The issue she focuses on is global warming. Her question is whether laypersons can reliably identify who to trust (and so what to believe) about global warming.

Empirical data suggests that, no matter how easy it may look to identify the genuine experts by applying Anderson's questions or Goldman's criteria, a lot of laypersons do not manage to reliably identify who the genuine experts about global warming are. Anderson's paper is from 2011, and she cites some then-contemporary data (from Newport 2010) to illustrate this point:

- In 2010, 48 per cent of the US public believed claims about the seriousness of global warming are exaggerated.
- In 2010, 50 per cent believed that human activity has caused global warming.
- In 2010, 50 per cent believed that most scientists think global warming is happening.
- From 2008 to 2010, the number of Americans 'dismissive' of global warming has more than doubled (to 16 per cent). The number 'alarmed' has halved (to 10 per cent).

At least in 2011, a very large percentage of the US population were not listening to climate scientists, who almost unanimously agree that global warming is happening and is caused by human activity (Cook et al. 2016).

It is at this point that the difference between Anderson's and Goldman's framing of the problem becomes important. Viewed from Goldman's perspective, the fact that some people seem either to not apply, or to do a bad job of applying, criteria like his or Anderson's is not really a problem. He is merely concerned to show that you *can* apply the criteria to form justified beliefs from expert testimony in situations where different (putative) experts say different things.

However, viewed from Anderson's perspective, the fact that some people do not apply them, or apply them badly, is a problem. Remember that, for her, the problem is, in part, political. We need to have democratic legitimacy for policy decisions that are informed by science, and we are not going to have it if large sections of the public reject the science that informs these policy decisions. Therefore, it is not enough for Anderson that laypersons *can* identify the genuine experts about something like global warming. It needs to be that most people *do* identify the genuine experts. What we need are some plausible strategies for getting them to identify the genuine experts and neither Anderson's questions nor Goldman's criteria are suitable to the task.

This leads Anderson to investigate why many people do not identify the genuine experts, at least when it comes to global warming. She draws on some empirical literature to give an explanation.

First, there is a lot of misinformation about global warming out there. Media reports about global warming are often misleading because they misrepresent the degree of scientific certainty (Dispensa and Brulle 2003). Further, the felt need on the part of media to present a 'balanced' account of issues by giving space for contrary opinions gives the misleading impression that the experts themselves are divided, and so may mislead the public. This isn't just idle speculation; Anderson cites various studies which find that the way the media reports global warming does indeed impact on public perceptions of the degree of scientific consensus (Nyhan and Reifler 2010; Skurnik, Yoon, and Schwarz 2005).

Second, we live in an increasingly segregated and partisan world. Anderson discusses Bill Bishop and Robert Cushing's 2008 book *The Big Sort*, which argues that Americans choose neighbourhoods, involve themselves in activities, and watch TV shows and channels that are compatible with their lifestyles and political views and values. Further, this process is self-perpetuating because of group polarization. If people only interact with others who agree with them on salient topics and issues, then their opinions tend to get firmer and more extreme over time. This exacerbates the trend towards homogenization (Talisse 2019).

Third, the partisan nature of American society is clear when you look at the differences between 'liberal' and 'conservative' views about global warming.[2] A 2016 report from the Pew Research Center (Funk and Kennedy 2016) found that:

- 70 per cent of 'liberal Democrats' trust climate scientists to give full and accurate information about the causes of global warming compared to only 15 per cent of 'conservative Republicans'.

[2] In this chapter, and throughout the book, I use these labels in the way they are used in US politics. It is fair to say that this reflects a US-centric framing of certain political issues. This framing is unfortunate, but it reflects the US-centric framing of the empirical literature I draw on in this and other chapters.

- 55 per cent of liberal Democrats say climate research reflects the best available evidence most of the time, and 39 per cent say it does so some of the time. In contrast, 54 per cent of liberal Republicans say this happens some of the time, and only 9 per cent say this happens most of the time.
- 57 per cent of conservative Republicans say climate research findings are influenced by scientists' desire to advance their career and 54 per cent say they are influenced by scientists' political leanings. In contrast, few liberal Democrats think either influence is common (16 per cent and 11 per cent, respectively).

Liberals are, on the whole, of the view that climate scientists are trustworthy sources of information about what is happening to the Earth's climate while conservatives are, on the whole, of the view that climate scientists are untrustworthy. Anderson explains this difference using empirical work on politically motivated reasoning (on which more in Chapters 4, 5, and 8). In particular, she explains it using work by Dan Kahan and collaborators (Kahan et al. 2011; Kahan, Jenkins-Smith, and Braman 2011; Kahan 2013, 2014).

Kahan and his collaborators explain partisan divergence over the existence of global warming in the following way (note that I am simplifying a bit here!). Conservatives recognize—if only implicitly—the tension between what climate scientists and policymakers say needs to be done to avert climate catastrophe and their deep distrust of state regulation of industry. They resolve this tension by downgrading their assessment of the degree of expertise of climate scientists. In contrast, liberals have no tension to resolve because there is an alignment between the sorts of policies recommended to combat global warming and the sorts of policies they are already inclined to support, such as state regulation of industry.

Anderson concludes that, due to (i) the ubiquity of misinformation about global warming, (ii) the increasingly partisan and segregated nature of social networks, and (iii) our tendency to engage in politically motivated reasoning, it is no surprise that many people fail to apply, or do a bad job of correctly applying, criteria for identifying who to listen to, and what to think, about global warming.

Now it is a jump from this to the conclusion that neither Goldman's criteria nor Anderson's questions are of any use in solving the general problem of identifying experts. But, while Anderson does not try to close this gap, we can say the following on her behalf. There is a lot of misinformation concerning a range of issues at the interface of science and public policy, such as the safety of nuclear power, the safety of vaccines, the risks posed by new technologies, and the risk posed by various viruses. Similarly, debate about these issues is very partisan, and this can be explained via similar mechanisms as those used to explain the partisan nature of the debate over global warming (Kahan et al. 2009, 2010; Kahan, Jenkins-Smith, and Braman 2011; Kahan, Hoffman, et al. 2012). It is therefore plausible to hold that Anderson has made a case for thinking that neither Goldman's criteria nor her questions are going to be much use in solving the general problem, not just the problem of identifying who to listen to about global warming.

If Anderson is right about this, what can be done about the problem of identifying experts? For Anderson, the lesson we should draw is that the problem of identifying experts cannot be solved by identifying criteria which individuals can apply. It may be that a few of us can make use of her questions or Goldman's criteria to reliably distinguish genuine from pseudo experts. But a lot of us are going to get things wrong a lot of the time. Given that Anderson sees the problem in partly political terms, she needs a solution that will work for the many, not just the few. She therefore urges that we adopt what she calls elsewhere an 'institutional' approach to the problem (Anderson 2006). The task for an institutional approach to the problem of identifying experts is to work towards constructing an epistemic environment in which it is easier for laypersons to identify the genuine experts and distinguish them from the pseudo experts. As Anderson puts it:

> creating an epistemically responsible democracy may require transforming social conditions so that ordinary citizens are disposed to reliably exercise their capacities for assessing expert trustworthiness.   (2011, p. 145)

The lesson we should draw is that we need an institutional epistemology. But, as we saw in Chapter 1, institutional epistemology comes in both ideal and non-ideal varieties. On a more ideal approach, the aim is to construct idealized versions of our socio-epistemic institutions and then measure actual institutions against their idealized versions. On a non-ideal approach, the aim is to make concrete suggestions for how to improve our epistemic environment. Anderson clearly favours the non-ideal over the ideal approach to institutional epistemology. But how might we go about improving our epistemic environment? She canvasses some suggestions:

- We can tackle the prevalence and influence of misleading media reports by revising norms of journalistic practice. For example, we can rethink the requirement to provide a 'balanced' overview of debates, at least when there are not equally good arguments on each side.
- We can encourage more dialogue and discussion across political divides.
- We can try to 'work round' politically motivated reasoning. There is valuable empirical work on how to frame the issue of global warming in ways that are more likely to secure cross-partisan acceptance of policies to mitigate its effects and address its root causes (I discuss this work in §4.2).

While none of these strategies are 'silver bullets', they show how we might get started in building a better epistemic environment—an environment where, while we may not be able to neuter cognitive biases, we can at least minimize their impact.

This completes my overview of Anderson's approach the problem of identifying experts. In the next section, I try to explain why there is a conflict between Goldman's and Anderson's respective approaches to this problem. But let me just reiterate the main respect in which their approaches differ. Where Goldman offers an individualistic solution to the problem of identifying experts, Anderson offers an institutional solution. Further, the institutional solution she offers is non-ideal, and so is a good example of a non-ideal approach to institutional epistemology.

3.4 Goldman vs. Anderson

So far, I have merely highlighted the differences between Goldman's and Anderson's approaches. In this section, I argue that, while there is a sense in which their approaches are compatible, there is also a sense in which they conflict. Further, I argue that, given how Goldman himself seems to understand the problem of identifying experts, Anderson's approach is superior to his own.

Let us start with the sense in which the two approaches are compatible. In §3.1, I said that Goldman's aims are modest. He wants to show that laypersons *can* apply his criteria to decide who (and who not) to trust. We should attribute a further, normative, claim to Goldman. Insofar as laypersons *can* apply these criteria in particular cases, they *should* apply them in these cases, at least if they want to obtain information about the matter at hand. We can view Goldman as making a modest claim within inquiry epistemology. There is a problem that many of us face in our inquiries (i.e. identifying experts) and there are situations where we can, and indeed should, apply his criteria to solve this problem. If we apply these criteria properly, then the result may be justified beliefs based on expert testimony.

It should be clear that none of the empirical considerations discussed in §3.3 provide reasons to reject this modest claim. That many laypersons do not apply Goldman's criteria due to politically motivated reasoning and the like is no argument against the claim that there are situations where laypersons should apply his criteria or against the claim that, when they do so properly, the result may be justified testimonial beliefs. So, if we understand Goldman as merely engaged in this modest project, there is no tension between his approach and Anderson's.

My aim in this chapter is not really Goldman exegesis, but it is a bit of a stretch to read Goldman as merely engaged in this modest project. Here are some relevant passages from Goldman (2001):

> some issues in epistemology are both theoretically interesting and practically quite pressing. That holds of the problem to be discussed here: how laypersons should evaluate the testimony of experts and decide which of two or more rival

experts is most credible. It is of practical importance because in a complex, highly specialized world people are constantly confronted with situations in which, as comparative novices (or even ignoramuses), they must turn to putative experts for intellectual guidance or assistance.   (p. 85)

The present topic departs from traditional epistemology and philosophy of science in another respect as well. These fields typically consider the prospects for knowledge acquisition in "ideal" situations. For example, epistemic agents are often examined who have unlimited logical competence and no significant limits on their investigational resources. In the present problem, by contrast, we focus on agents with stipulated epistemic constraints and ask what they might attain while subject to those constraints.   (p. 85)

And finally:

There is no denying, however, that the epistemic situations facing novices are often daunting. There are interesting theoretical questions in the analysis of such situations, and they pose interesting practical challenges for "applied" social epistemology. What kinds of education, for example, could substantially improve the ability of novices to appraise expertise, and what kinds of communicational intermediaries might help make the novice-expert relationship more one of justified credence than blind trust.   (p. 109)

We can draw on these passages to better understand how Goldman sees his own project. First, he sees the problem of identifying experts as practical, not theoretical. Second, he thinks it calls for a non-ideal approach—he wants to consider a problem faced by real epistemic agents, not imaginary epistemic agents. Third, and bringing these two things together, he sees the problem as belonging within the remit of 'applied epistemology'. While this is a less familiar term than 'applied ethics', it can be understood in a parallel way. Applied ethics looks to apply ethical theories and considerations to real-world problems and to examine these problems in their full complexity, without recourse to simplifying assumptions and idealizations.

Similarly, applied epistemology seeks to apply epistemological theories and considerations to real-world problems and to examine these problems in their full complexity, without recourse to simplifying assumptions and idealizations (Coady and Chase 2018). Applied epistemology conceives of itself as addressing practical problems and it eschews (or at least tries to eschew) idealizations in doing so. It is therefore closely related to non-ideal epistemology as I understand it. (Closely related, but still different. I am not entirely confident that Chapters 6, 7, and 8 really qualify as applied epistemology.)

While it is hard to square Goldman's methodological remarks in his paper with the modest project, it is even harder to square his published views on the aims of social epistemology with the modest project. In 'Why Social Epistemology is *Real* Epistemology' Goldman puts his broad project as follows:

> Large chunks of analytic philosophy, epistemology included, focus on everyday thought and talk, often teasing out norms that implicitly govern them. Epistemology should not be restricted, however, to this activity. A broader view of epistemology is present in the epistemology of science, which definitely does not assume that lay practices are the exclusive or final benchmark of anything normative. Philosophers of science are more sympathetic to the meliorative project of improving epistemic practices, especially in the scientific arena. In addition to science, there are other arenas where meliorative epistemological projects make eminently good sense. Many sectors of social life feature practices and institutions ostensibly dedicated to epistemic ends, but where one is entitled to wonder whether prevailing practices and institutions are optimal. Subjecting such practices and institutions to epistemic evaluation is therefore in order. Are they the best practices or institutions (of their type) that can be devised? What alternative practices would work better in epistemic terms? This kind of social epistemological project was advocated in *Knowledge in a Social World* [Goldman's major contribution to social epistemology].   (2010b, p. 18)

I do not really see a difference between the sort of meliorative approach Goldman is describing and Anderson's 'institutional' approach to the problem of identifying experts. We have a social practice (layperson deference to experts), and our question is how that practice can be improved to better serve an epistemic end (laypersons having true beliefs about important issues). Anderson thinks the empirical evidence tells us that the way to change the social practice to better serve this epistemic end is not to try and get individuals to apply different or better criteria for identifying experts but to construct a better epistemic environment. Whether she is right about this or not, it is hard to see how Anderson could be addressing a different question to the one these remarks suggest Goldman wants to address.

All told, it is tempting to view Goldman as *intending* to do something like what I am calling non-ideal institutional epistemology but failing because he still relies on too many idealizations. The question of whether to count a worthy but misguided *attempt* at doing non-ideal institutional epistemology as an example of ideal or non-ideal epistemology does not strike me as particularly important, so I will set it to one side. The important thing is that, by his own lights, Goldman should regard the fact that most laypersons will *not* do a good job of identifying genuine experts if they try to apply his criteria as an objection to those criteria.

Let us leave the Goldman exegesis behind for a minute and consider the modest project on its own merits. What about someone who defends the modest project?

Can't they just shrug and say that Anderson's approach isn't preferable to theirs, simply different? Well, yes, they can. But there are two points I want to make here.

First, just how 'modest' is the modest project? If it is too modest—if the claim is just that there are certain (quite contrived) situations in which criteria like Goldman's can be successfully applied—then it is hard to see how it is a 'solution' to the problem of identifying experts. Consider an analogy. Responses to the perennial epistemological problem of scepticism about the external world are viewed as adequate only if they can vindicate our sense that we know quite a lot about the external world. It is accepted that it is not enough for them to show that we have *some* knowledge, still less that we *could* have knowledge. They need to show that we know a lot—perhaps even as much as we ordinarily take ourselves to know (DeRose 1995; Pritchard 2002; Sosa 1999).

Similarly, you might think that a viable solution to the problem of identifying experts needs to do more than show that we *can* identify experts. It needs to show that we can do this in a wide range of situations. (This is not to say that it needs to show how we can do it in all situations.) It is arguable that Goldman's strategy of showing that there are contrived situations in which the criteria can be successfully applied only succeeds in establishing the weaker claim, not the stronger and more interesting claim. What needs to be shown is that these criteria are useful in a wide range of situations. But this is precisely what Anderson is arguing cannot be done.

Second, this is a nice illustration of the more general point that what often lies at the heart of debates between ideal and non-ideal theory—whether in epistemology or elsewhere—is a question about which projects we should pursue. You *can* pursue the modest project. Indeed, it seems clear that there are some good reasons for doing so. It is important to know whether it is possible to form justified beliefs based on expert testimony, particularly if the phenomenon of expert disagreement casts doubt on whether this is possible. But if you have any aspirations to doing epistemology that has real-world relevance—as Goldman clearly does—we need to do a lot more than show it is possible to have justified beliefs based on expert testimony. Of course, you might have no aspirations to doing epistemology that has real-world relevance. But if your project is not meant to be relevant then it is important to be clear about this from the outset.

3.5 Two Tasks for Anderson

In this chapter, I have defended Anderson's non-ideal institutional approach to the problem of identifying experts. In broad outline, Anderson re-frames the problem. Rather than, as Goldman does, asking what I, as an individual inquirer, can do to identify experts, she asks how we can construct systems of knowledge production and dissemination that do a better job of producing and disseminating knowledge. Anderson's thought is that the polluted epistemic environment in which we operate, the highly partisan nature of social networks, and the cognitive biases that influence our thinking mean that, no matter how sensible the criteria you propose, laypersons are often going to struggle to identify genuine experts. The only way round this is to try to minimize the impact of these distorting forces by making it easier for laypersons to defer to the genuine experts and ignore the pseudo experts. The only way to solve the problem is by building a better epistemic environment.

There are, however, still two tasks that need to be carried out to vindicate Anderson's approach. First, it is one thing to make some suggestions about how we might construct a better epistemic environment and quite another to provide concrete evidence that these suggestions will work in practice. A non-ideal institutional epistemology aims at the latter, not the former. More needs to be said about how a better epistemic environment is to be constructed. I take up this task in the following chapter. In doing so, I argue for what you might call an 'evidence-based' approach to non-ideal institutional epistemology.

Second, there is a fundamental objection to Anderson's approach that must be dealt with. Recall that Anderson's original motivation is to secure democratic legitimacy for science-based public policy. But when she shifts to discussing how we might build a better epistemic environment she does not discuss whether the suggestions she makes are themselves democratically legitimate. Perhaps her thought is that, if these suggestions worked, they would be to everyone's benefit, especially the benefit of those whose epistemic positions would be improved as a result. If that is her thought, then what she is proposing sounds a lot like paternalism. But a lot of people, philosophers included, are very suspicious of paternalism.

This is not to say that there are not isolated cases where paternalistic interference might be justified; there surely are. But there is a standing presumption against paternalism because it seems to infringe on our autonomy. Paternalism is therefore to be viewed as a last resort—as something to be used when there is no other alternative, not as a general strategy. To vindicate Anderson's approach, and non-ideal institutional epistemology in general, I need to either defend the permissibility of certain forms of paternalistic interference or show that the forms of interference Anderson is proposing are not really paternalistic. I take up this task in the two chapters that follow. Taken together, then, these chapters are a defence of a non-ideal approach to institutional epistemology.

4

Persuasion and Paternalism

We all hold some false beliefs. Many of these false beliefs are inconsequential. My false belief that Tallinn is the capital of Latvia is of little consequence. Sadly, other false beliefs are more consequential. Many of us have false beliefs about the causes and reality of global warming, the safety of nuclear power, genetically modified organisms, or vaccines. In this chapter, I discuss the problem of what to do about consequential false beliefs. My central question is what, if anything, we can do to stop people having false beliefs about these sorts of issues.

The reader might be wondering how this problem relates to the problem of identifying experts. The problems themselves are related because false beliefs about who the experts are will often be consequential. But the more important connection is that my discussion of the problem of consequential false beliefs will end up addressing the two tasks I said need to be carried out to vindicate Anderson's approach to the problem of identifying experts.

The first task was to say more about how a better epistemic environment might be constructed. This task is important because one of the distinctive features of a non-ideal approach to institutional epistemology is that its suggestions for how to construct a better epistemic environment are 'evidence-based'—they are based on the best available empirical evidence. It is therefore incumbent on the non-ideal institutional epistemologist to produce the relevant evidence. The second task was to address the worry that institutional epistemology is committed to a problematic form of paternalism. In this chapter, I take up both tasks, though in the context of the problem of consequential false beliefs. This chapter therefore continues my defence of non-ideal institutional epistemology.

In addressing the problem of consequential false beliefs, it is important to distinguish two issues. The first, which is empirical, concerns which methods are *effective* in persuading people that they hold false views about issues like global warming. In answering the first question, we can draw on multiple empirical literatures, including work on (the science of) science communication, the psychology of persuasion, and motivated reasoning. It is from these empirical literatures that we get the evidence on which non-ideal institutional epistemology must be based. In addressing this issue, then, I hope to complete the first task that needs to be carried out to vindicate a non-ideal approach to institutional epistemology.

The second issue, which is normative, concerns which methods we are *permitted* to use in the service of persuading people to change their minds. There are two ways in which you might frame the normative concern, one of which is ethical, the

Non-Ideal Epistemology. Robin McKenna, Oxford University Press. © Robin McKenna 2023.
DOI: 10.1093/oso/9780192888822.003.0004

other of which is more political. The ethical framing is a little less involved, so let us start with that. It is clear that some ways of trying to persuade people to change their minds are unethical. While we can debate the effectiveness of brainwashing, it is clear that it is ethically dubious, to say the least. Even if we *could* 'solve' the problem of consequential false beliefs by an extensive programme of brainwashing, that does not mean that we *should* try to solve the problem in this way. There is a gap between showing that some method for solving the problem will be effective and showing that we are permitted to use that method.

The political way of framing the problem is related to the way Anderson frames the problem of identifying experts. Anderson identifies a tension between the requirements of responsible public policymaking and democratic legitimacy. Responsible public policymaking in a technologically advanced society should be based on the best available scientific evidence. But to be democratically legitimate there must be broad (though of course not universal) acceptance of the policies which are put in place. This, in turn, seems to require broad acceptance of the science on which the policies are based.

It is here that the tension arises. There are many public policy issues where sizeable minorities of people reject the science on which responsible public policymaking might be based. Consider, for instance, global warming. In many countries, including the USA, there is a sizeable minority of the population who rejects the science on global warming (Ballew et al. 2019; Tranter and Booth 2015). This minority tend to hold similar political views (broadly, pro-free market and anti-regulation) and support the same political parties (Ballew et al. 2019). Once politics and science become intermingled in the way they are in debates about global warming, the prospects of any broad consensus emerging seem dim, given the difficulty of persuading people to change their minds about things they regard as integral to their political and cultural identities.

While these two ways of framing the problem are distinct, I will run them together. This is because they share a key idea, and it is really this key idea that will be my primary focus in this and the following chapter. Many hold that some ways of trying to persuade people to change their minds are problematic because they infringe on their *intellectual autonomy*. The thought is that some methods of persuasion seek to bypass our capacity for critical reflection and deliberation, whether in an overt way (as in brainwashing) or in a more subtle way (as in, for example, 'nudges' in the sense of Thaler and Sunstein 2008). The problem with methods of persuasion that infringe on intellectual autonomy can be put in ethical terms (intellectual autonomy is valuable so there is a strong presumption against infringing on it) or political terms (infringing on someone's intellectual autonomy is a sort of coercion and coercion is not democratically legitimate). Either way, the key idea is that we need to distinguish between permissible and impermissible methods of persuasion. Permissible methods respect intellectual autonomy; impermissible methods do not.

I am going to argue (in Chapter 5) that the attraction of intellectual autonomy as an epistemic ideal depends on certain idealistic assumptions about human cognition and our epistemic environments. While intellectual autonomy may be an attractive ideal from the perspective of ideal epistemology, its value is more dubious from the perspective of non-ideal epistemology. This, in turn, casts doubt on arguments against methods of persuasion based on the importance of protecting intellectual autonomy.

My aim in this chapter is more modest. Empirical work shows that, often, the most effective methods of persuading people to change their minds about politically contentious scientific issues are what we might call 'marketing methods'— methods that seek to 'sell' science, rather than persuade purely through the force of reason and evidence. I will, first, explain why many are worried about such methods on the grounds of intellectual autonomy. I will then explain why these worries are misplaced. I leave the question of the value of intellectual autonomy itself for the next chapter. Taken together, this chapter and the next will complete the second task that needs to be carried out to defend a non-ideal approach to institutional epistemology—they defend it against the charge that it is objectionably paternalistic. (It may be paternalistic, but it is not objectionably so.)

Here is the plan. I start by saying a little more about the non-ideal approach to the problem of consequential false belief that I adopt in this chapter (§4.1). I then review some empirical work on which methods are most effective in persuading people to change their minds about issues like global warming (§4.2). It turns out that many of the most effective methods involve trying to structure the epistemic environment in such a way that epistemic agents are more likely to form true beliefs about the relevant scientific issues than they would otherwise be. They are 'science marketing' methods. Next, I make the worry that such methods infringe on our intellectual autonomy more precise (§4.3). I argue that these methods are tantamount to an epistemic form of paternalism, and paternalism is generally thought to infringe on autonomy (Dworkin 2020). After making this objection more precise, I address it by arguing, first, that these science marketing methods need not infringe on our intellectual autonomy (§§4.4–4.5) and, second, that it is harder to draw a neat distinction between methods of rational persuasion and marketing methods than many suppose (§4.6). I finish by raising the issue that I focus on in the next chapter, which is the value of intellectual autonomy itself (§4.7).

4.1 Non-Ideal Institutional Epistemology

What does epistemology have to contribute to the problem of consequential false belief? On one approach, the epistemologist's task is to identify norms of inquiry that an individual should follow to navigate complicated political and scientific issues. These norms might tell the individual how to gather evidence, who to ask for information, or how to identify who to trust. The idea would be that, if they follow these norms, the individual is likely to form true beliefs and avoid forming false ones. This is the approach that we saw Goldman takes to the problem of identifying experts. On another approach, the epistemologist's task is to suggest how we might improve our systems of knowledge production and dissemination and more broadly our epistemic environment. This is the approach that we saw Anderson takes to the problem of identifying experts. In this and the following chapter, I adopt the second, institutional approach to the problem of consequential false belief. But let me start by saying more about this approach. What follows is my attempt to draw out a methodology that is implicit in Anderson's work (Anderson 2006, 2011). For a more detailed and sustained defence of a similar methodology as applied to scientific testimony, see Gerken (2022).

Institutional epistemology has *ameliorative* aims. The task is to figure out how to improve our epistemic environment. Of course, what counts as a 'better' epistemic environment depends on your epistemological views. Some hold that what matters is the prevalence of true belief (Goldman 1999). Others might suggest that what matters is the prevalence of knowledge. This would be a 'knowledge first' approach to institutional epistemology in the manner of Williamson (2000). I will set the question of what makes for a better epistemic environment aside as it is really a question for the theorist of epistemic ideals. For simplicity, though, I will usually talk as if the first, 'veritistic', view is right. In places, though, I will talk as if the second, knowledge-based view, is right. (I already did this in Chapter 3.)

We can distinguish between two ways of going about institutional epistemology. One way would be to start with some ideas about what good systems of knowledge production and dissemination or a good epistemic environment would look like. You might start with a simple, idealized model of a social institution (like science) on which it does a good job of producing and disseminating knowledge. You would then measure up our actual institutions and systems against this ideal. To the extent that they measure up well, we can say that these institutions and systems are working well; to the extent that they measure up badly, we cannot. So, on this approach, you start by forming a picture of an ideal and then ask to what extent our actual systems approximate that ideal. This is what we might call an ideal approach to institutional epistemology. Philip Kitcher's idea of 'well-ordered science' is a good example of the ideal approach (Kitcher 2001, 2011).

Another way of approaching institutional epistemology would be to start with our existing systems of knowledge production and dissemination. When they work well (when they produce and disseminate knowledge), how do they do it? When they work badly (when they do not produce or do not disseminate knowledge), what goes wrong? This is what we can call an *evidence-based* or non-ideal approach to institutional epistemology. On this approach, the task is to, first, gain an understanding of how our existing systems of knowledge production and dissemination work and then, second, to identify concrete ways in which they could be improved to do a better job of producing and disseminating knowledge. In this chapter, I tackle the problem of consequential false belief using an evidence-based, non-ideal approach.

There is a large empirical literature on how people form beliefs about prominent issues like global warming, and there is also a large literature on which interventions are likely to correct misperceptions about these sorts of issues. On the evidence-based or non-ideal approach to institutional epistemology, these literatures directly inform our suggestions for improving our systems of knowledge production. Suggestions that might seem plausible in the abstract may not prove so effective in the 'real world'. If our aims are genuinely ameliorative, it is no good producing a list of prescriptions that work in the abstract but not in practice. If they are to be adequate to the task, our prescriptions for how to solve the problem of consequential false beliefs need to build on relevant work on the psychology of persuasion, attitude-change, and science communication.

You might worry that this non-ideal approach stands to epistemology as optometry stands to epistemology. An optometrist knows which glasses will improve vision and so can tell someone what to do to improve their perceptual belief-forming processes (wear these glasses!). Similarly, it may be that by drawing on empirical literature we can figure out how to improve the epistemic environment and enable people to have more knowledge or true beliefs. But, just as an optometrist is not doing epistemology when they figure out which glasses will work best for their customer (or when they recommend that their customer wear those glasses), you are not doing epistemology when you figure out how to get people to change their minds about issues like global warming. To be sure, it may be that the question of how to get people to change their minds about such issues is important (optometry is important!). But it is not an *epistemological* question. It is a question for science communicators, sociologists, educators, and so on, not for epistemologists.

I largely agree that merely identifying empirically supported strategies for changing minds is not doing *epistemology*. (I say 'largely' because one thing an epistemologist might add is an appreciation for distinctions like that between belief and justified belief. Some ways of changing minds might yield justified beliefs while others yield beliefs but not justified beliefs.) But I do more than identify some strategies. I also address recognizably *philosophical* issues that would arise from putting these strategies into practice. These issues include whether these strategies infringe on intellectual autonomy or constitute a problematic form of epistemic paternalism. These issues belong within epistemology, at least if you construe epistemology as involving more than the theory of knowledge or justification, as I do.

More generally, my approach in this chapter, and this book, belongs to a tradition within epistemology that is concerned with improving our epistemic practices—what I have referred to as 'inquiry epistemology'. What I am doing is therefore not analogous to optometry, though it may be that, along the way, I rely on some empirical literature that stands to socially embedded processes of belief formation in much the same way that optometry stands to our perceptual belief-forming systems.

4.2 Gathering the Evidence

How do we form views about issues like global warming? How might misperceptions about it be dispelled? In this section, I give an overview of some empirical research on these questions. As it will turn out, the result of this survey will be that many of the most effective methods for tackling the problem are 'marketing methods'—they try to sell science rather than assuming it sells itself.

Let us start with some figures: 97 per cent of climate scientists agree that human activity is a major cause of global warming (Cook et al. 2016). But a recent study in the USA found the following (Ballew et al. 2019):

- Around 7 in 10 Americans think global warming is happening and around 1 in 8 Americans think global warming is not happening.
- Around 6 in 10 Americans think that global warming is mostly human caused and 3 in 10 think it is due mostly to natural changes in the environment.
- Just over 50 per cent of Americans realize that most scientists think global warming is happening. But only about 1 in 5 realize how strong the level of consensus among scientists is.[1]

The good news is that 7 in 10 Americans think global warming is happening. The slightly less good news is that only 6 in 10 think it is mostly human caused. The more unwelcome news is that only around 50 per cent of Americans realize that scientists are almost unanimous in agreeing that global warming is happening.

To ascertain how good or bad this news is we really need to look at *who* thinks these things. It turns out that views about global warming (whether it is happening, what is causing it, and whether scientists agree about it) correlate strongly

---

[1] For some cross-national data, see Tranter and Booth (2015). While some countries do better than others (and the USA does badly), there is no country where most people have accurate beliefs about the level of scientific consensus.

with political views. Ballew et al. found that almost all 'liberal Democrats' think that global warming is happening, as do most 'moderate Democrats'. In contrast, fewer than 50 per cent of 'conservative Republicans' think global warming is happening. It is reasonable to suppose that even fewer conservative Republicans think that global warming (if it is happening at all) is mostly human caused, or that most scientists think global warming is happening. In the USA, global warming—or the recognition of its reality—poses a political problem. While there is broad acceptance that global warming is happening, there is a sizeable minority, which happens to be politically powerful, that denies this. The question is: how could we persuade them that they are wrong?

There are lots of things you might try to persuade people that they are wrong about something like global warming. Perhaps the problem is a general lack of scientific understanding. If so, the solution must involve more and better science education, perhaps allied with a greater emphasis on critical thinking in education (Bak 2001; Sturgis and Allum 2004). There is ample evidence that providing people with relevant facts and trying to fix gaps in their understanding can be effective. But there is also ample evidence that, at least in the case of global warming, merely providing the relevant facts has not been particularly effective (Downing and Ballantyne 2007; Gardner and Stern 1996; Moser and Dilling 2011). As Moser and Dilling put it:

Clearly, much can be said for broad public education in the principles and methods of science in general and in climate science specifically. A sturdier stand in science education may leave lay individuals less susceptible to misleading, factually untrue argumentation. But ignorance about the details of global warming is NOT what prevents greater concern and action.   (2011, p. 163)

If the problem were a lack of scientific understanding, you would expect that how likely someone is to accept that global warming is real would be correlated with their level of scientific understanding (the higher your level of scientific understanding, the more likely you are to accept it is real). But this is not the case. In fact, the strongest influence on views about global warming seems to be political ideology, not level of scientific understanding or literacy (Hamilton 2011; Hamilton et al. 2015; Hardisty, Johnson, and Weber 2010; Hornsey et al. 2016; Kahan, Jenkins-Smith, and Braman 2011; Lewandowsky and Oberauer 2016; Tranter and Booth 2015). For instance, a 2019 Pew Research Center survey tells us that, as you would expect, the percentage of liberal Democrats who think that global warming is mostly human caused increases with level of scientific understanding. But conservative Republicans who are better informed about the underlying science are slightly *less* likely to think that global warming is mostly human caused than conservative Republicans who are less well informed (Funk and Kennedy 2016).

What is going on here? There are two key reasons why providing people with relevant facts might not lead them to change their minds about an issue like global warming and why views about the issue correlate more with political ideology than level of scientific understanding. The first is the influence of politically motivated reasoning, which was briefly discussed in Chapter 3. As we saw, we do not process and evaluate the information we receive in a neutral or unbiased way. Rather, our background beliefs, views, and values—including our political beliefs, views, and values—influence our information processing (Kahan, Jenkins-Smith, and Braman 2011; Lewandowsky and Oberauer 2016). Thus, liberals tend to form a more favourable view of the scientific evidence on global warming than conservatives. Conservatives have a motive to reject the science on which policies to tackle global warming are based because these policies clash with their basic political conviction that there should be limited regulation on industry. In contrast, because there is no tension between the steps that need to be taken to deal with global warming and liberal political inclinations, liberals have no motive to reject the science on global warming.

It is important to note that *everyone* underestimates the extent of scientific consensus on global warming. While liberals tend to think the level of scientific consensus on global warming is higher than conservatives think it is, they still underestimate the true level (Ballew et al. 2019). This means that politically motivated reasoning can only be part of the story.

The second part, which was also briefly discussed in Chapter 3, has to do with the prevalence of misinformation about global warming in the public sphere (Cook 2016, 2017; Cook et al. 2018). Everyone is affected by the misinformation spread by global warming deniers because (put roughly) misinformation tends to 'drown out' information. Informing the public that scientists agree on certain issues (such as global warming) can be very effective in increasing public acceptance, even among those who, for ideological reasons, are disposed to deny the existence of consensus on global warming (Bolsen, Leeper, and Shapiro 2014; Cook and Lewandowsky 2011; Lewandowsky, Gignac, and Vaughan 2013; van der Linden et al. 2014). But this effect is drastically reduced when there is widespread misinformation about the level of scientific consensus and the integrity of climate scientists (Cook 2016, 2017; van der Linden et al. 2017). As van der Linden et al. put it:

> Results indicate that the positive influence of the 'consensus message' is largely negated when presented alongside [misinformation]. Thus, in evaluating the efficacy of consensus messaging, scholars should recognize the potent role of misinformation in undermining real-world attempts to convey the scientific consensus.   (2017, p. 5)

The problem is not just that the public do not understand climate science or are scientifically illiterate. Any lack of understanding or scientific illiteracy is exacerbated by, first, the ubiquity of politically motivated reasoning and, second, the prevalence of misinformation. If this is right, then strategies for dealing with public misperceptions that assume the task is to correct misunderstandings by supplying relevant facts are not going to solve the problem.

What can we do? I am going to give brief overviews of three strategies that have been proposed in the literature on climate science communication and outline some of the evidence supporting them. In each case, the strategy is a marketing method—it aims to sell a product (climate science) rather than to persuade by (merely) supplying relevant information or argumentation.

First, we can consider how issues like global warming are *framed*. There is a large body of evidence suggesting that whether individuals are willing to accept specific global warming mitigation policies—and even the science underpinning them—depends on how those policies are described (Campbell and Kay 2014; Corner et al. 2015; Dahlstrom 2014; Dryzek and Lo 2015; Hardisty, Johnson, and Weber 2010; Kahan 2014; MacInnis et al. 2015). For instance, discussions of what to do about global warming are often framed in terms of what we can do to reduce carbon emissions. This leads to a situation where (crudely) conservatives need to choose between the science and their conviction that business should be free from government interference. There is evidence that framing the problem as amenable to technological solutions (e.g. geoengineering) can make those who are ideo-logically opposed to regulating carbon emissions more willing to accept that action is needed to combat global warming (Kahan et al. 2015). To take another example, there is evidence that framing charges for environmental costs as 'carbon offsetting' rather than as a 'carbon tax' increases public acceptance of the necessity of the charges (Hardisty, Johnson, and Weber, 2010).

Second, as well as considering the content of the message, we can look at who delivers it. Climate scientists have generally been the messengers of choice, which makes sense because public trust in scientists is quite high (Ipsos MORI 2014; American Academy of Arts & Sciences 2018). But scientists are not the most trusted source on every issue, or always best placed to communicate key messages on a politically contentious issue such as global warming (Cvetkovich and Earle 1995; Cvetkovich and Löfstedt 1999; Kahan et al. 2010; Moser and Dilling 2011). When it comes to politically contentious issues, we tend to rely on those whom we think share our social and political values. Put crudely, we would rather listen to someone who is 'like us' than someone who isn't, even if they are less likely to be right (Kahan, Jenkins-Smith, and Braman 2011). The upside of this is that, if we take steps to ensure that a politically diverse group makes the case for action on global warming, we have reason to think that this would be effective.

Third, and finally, we can consider which persuasive strategies we use in the first place. It is common to pursue a 'debunking' strategy: take some claim made by global warming sceptics, or others who want to frustrate efforts to tackle the problem of global warming, and then show why the claim is false. Debunking clearly can work, but it is often less effective than you might expect because misconceptions and false beliefs are surprisingly hard to dislodge (Lewandowsky et al. 2012; Seifert 2002). For this reason, many have considered an alternative strategy called 'prebunking'. Where debunking is a matter of refuting a belief that has already been accepted, prebunking is a matter of giving people the tools to avoid being taken in by false claims and misinformation in the first place.

Prebunking is based on an intuitively plausible idea from 'inoculation theory' (Compton 2013). The thought is that you can 'inoculate' someone against certain forms of misinformation and misleading ways of arguing by exposing them to these forms of misinformation and misleading ways of arguing in a controlled environment (Bolsen and Druckman 2015; Cook, Lewandowsky, and Ecker 2017; Ivanov et al. 2015; Pfau and Burgoon 1988; Pfau 1995; van der Linden et al. 2017). For example, you might present someone with a common climate sceptic argument along with a refutation of it. Take the claim that human CO_2 emissions are tiny in magnitude compared to natural emissions. This claim is true. But it does not support the conclusion that human emissions are irrelevant because what human emissions primarily do is interfere with the natural carbon cycle, putting it out of balance. The idea is that you would accompany a presentation of this climate sceptic argument along with the explanation why it is a bad argument.

Let me sum up. In this section, I have discussed three strategies that we might use to persuade people to change their minds about issues like global warming. In each case, there is clear empirical evidence that the strategy in question is likely to be effective. They are good examples of the sorts of strategies for constructing a better epistemic environment that should be proposed by the non-ideal institutional epistemologist. While more could be said here, I take this to be a good first stab at drawing up a list of evidence-based suggestions for improving our epistemic environment and so to go some way towards completing the first task that needed to be carried out to defend a non-ideal approach to institutional epistemology.

In the rest of this chapter, I turn to the second task, which was to deal with the worry that non-ideal institutional epistemology (or rather the sorts of persuasive strategies it proposes) inevitably infringes on our intellectual autonomy. The methods of persuasion I have highlighted in this section involve targeting groups of people with a message or in a way that is designed to produce a certain outcome (they abandon a false belief and adopt a true one). In the next section, I look at why you might think marketing methods are problematic because they infringe on our right to make up our own minds about issues that matter to us.

4.3 Epistemic Paternalism and Intellectual Autonomy

The fact that a strategy is likely to be effective is not conclusive reason for deploying it. Maybe there is something ethically or politically problematic about the 'science marketing' methods I discussed in the previous section. In this section, I make this worry more precise, before addressing it in the following sections. My suggestion is that we can frame the worry in terms of *autonomy* and *paternalism*. Specifically, we can frame it in terms of *intellectual autonomy* and *epistemic paternalism*.

Let us start with what epistemic paternalism is. Put roughly, a practice is paternalistic if it involves interfering with someone's choices or actions for their own good, but without their consent (Dworkin 2020). If I hide your cigarettes because I know that otherwise you will smoke the whole packet, which will be bad for your health, then I act paternalistically towards you. Equally roughly, a practice is 'epistemically paternalistic' when it involves interfering with someone's cognitive activities—with the conduct of their inquiries—with the aim of improving their epistemic position, but without their consent (Ahlstrom-Vij 2013). For example, if a judge withholds information about the defendant's criminal record from the jury because they think knowing this would bias their deliberations and so make them less likely to arrive at the correct verdict, the judge engages in epistemic paternalism. He has interfered with the jury's cognitive activities (by withholding information) with the aim of making it more likely that they arrive at the correct verdict.[2]

---

[2] There are issues with Ahlstrom-Vij's definition of epistemic paternalism and with some of the examples he uses to illustrate it. I do not consider these issues in the main body of the text because it would take me too far away from my main purpose. My aim in this section is to articulate why some might be suspicious of the science marketing methods discussed in §4.2. I use the language of epistemic paternalism to this purpose because what I am describing fits with how Ahlstrom-Vij understands epistemic paternalism. That said, I want to comment on three issues. First, you might wonder what makes epistemic paternalism a distinctive form of paternalism. Isn't epistemic paternalism just common-or-garden paternalism (Bullock 2016)? I do not have a satisfactory answer to this question because I do not have a satisfactory answer to the more fundamental question of what makes something 'epistemic'. But perhaps we can say that epistemic paternalism is distinctively epistemic in roughly the same sense in which Fricker (2007) thinks epistemic injustice is distinctively epistemic. Just as an individual experiences epistemic justice when they are harmed in their capacity as an epistemic agent, an individual is interfered with in an epistemically paternalistic way when they are interfered with in their capacity as an epistemic agent. Second, in the jury example, the judge's withholding certain information from the jury only counts as (epistemically) paternalistic if the jury does not consent to this and the judge withholds the information for the jury's own (epistemic) good. You might question whether either condition is met. The consent condition may not be met because it could be that the jury is fully aware that some information is being withheld from them and have no objection to this being the case. The other condition may not be met because it could be that the judge is not really concerned with the jury's epistemic good or health. They are just concerned with ensuring a fair trial and they think that withholding some information from the jury is necessary for this end. Ahlstrom-Vij has a response to this. But it is not entirely convincing and involves getting involved in speculations about the motives underlying interventions like that of the judge. Finally, parallel worries can be raised about whether the science marketing methods I discuss in this chapter are really (epistemically) paternalistic. It is more plausible that the consent condition is met here. The effectiveness of science marketing may well require that its targets be unaware that they are being targeted by it (it is possible that its targets might object to it if they were aware). It is perhaps also more plausible that the second condition is met. You can see why science communicators might be interested in the public's 'epistemic health' as well as securing wider public support for certain policy interventions. But, to reiterate, I am not committed to the claim that these methods are examples of epistemic paternalism. What I am committed to is the claim I go on to defend, which is that they need not infringe on our intellectual autonomy. If they are not examples of paternalism, this helps rather than hinders my case here!

Many think that paternalistic interference is always prima facie objectionable because any form of paternalistic interference infringes on our right as autonomous individuals to make our own choices and our own decisions about how to act (Feinberg 1986; see also the essays in Grill and Hanna 2018). The point is not that any sort of infringement on our autonomy is prima facie objectionable. If I have autonomously decided to murder my neighbour, then there is nothing prima (or ultima) facie objectionable about your calling the police to stop me because you are concerned for my neighbour's safety. The point is that it is always prima facie objectionable to infringe on someone's autonomy *for their own good*.

An analogous objection can be made against epistemic forms of paternalism. The objection is that epistemically paternalistic interference is always prima facie objectionable because it infringes on our intellectual autonomy (Ahlstrom-Vij 2013, ch. 4). In our example, the judge's withholding information about the defendant's criminal record may be regarded as prima facie objectionable because it infringes on the right of the jury to make up their own minds about what they think. Whether it is ultima facie objectionable is, of course, another question. In this case it may be that there are good reasons for keeping the jury 'in the dark' when it comes to certain facts about the defendant. Those reasons are, at least in part, epistemic. Withholding this information may prevent the jury forming a biased and false impression of the likelihood that the defendant is guilty. Still, the point is that there is a standing presumption against epistemically paternalistic interference because it infringes on our intellectual autonomy.

It is arguable that the science marketing methods I discussed in §4.2 are also epistemically paternalistic forms of interference. They involve interfering with the right of individuals to conduct their inquiries in the way that they see fit with the aim of making them epistemically better off but without their consent. You might therefore think that, just like the judge's withholding of information from the jury, these methods infringe on our intellectual autonomy and so are prima facie problematic. To firm up this initial impression I want to compare these methods with nudging, in the sense of Richard Thaler and Cass Sunstein (2003, 2008).

According to Thaler and Sunstein, a nudge is 'any aspect of the choice architecture that alters behaviour in a predictable way without forbidding any options' (2008, p. 6). For example, imagine a doctor needs to inform their patient about the chance that a potentially lifesaving but risky operation will be successful. This doctor has a choice to make in how they frame the decision. Either they can say what percentage of patients who have the operation are alive in x years' time or they can say what percentage of patients who have it are dead in x years' time. Of course, these ways of presenting the decision are statistically equivalent. But there is evidence that focusing on how many patients live makes it more likely that the patient will opt to have the operation than focusing on how many patients die (McNeil et al. 1982). A doctor who knows this and thinks it is in the patient's best interests to have the operation might present the odds in the way most likely to lead to the patient choosing to have the operation. If the doctor does this, the patient is not forced to choose to have the operation. They can always still refuse. But the doctor knows that, in framing the choice this way, the chances of the patient agreeing increase.

This example illustrates the crucial aspect of nudging which is that it involves interfering with someone's decision-making with the aim of making them better off without their consent yet also without any sort of coercion. This is why Thaler and Sunstein describe nudging as a 'libertarian' form of paternalism. Nudges are paternalistic in that they involve interfering with someone's decision-making with the aim of making them better off without their consent. They are libertarian in that they do not reduce freedom of choice (the patient in the example is free to decline to have the operation).

There is a debate about whether nudges are really libertarian (Mitchell 2005). There is also a debate about whether some of Thaler and Sunstein's central examples are really examples of paternalism (Dworkin 2020). Rather than get embroiled in these issues I want to focus on the connection between nudging and autonomy. Nudges are meant to not infringe on freedom of choice. But some have objected that nudges can still infringe on autonomy even though they do not reduce the number of available options. As Daniel Hausman and Brynn Welch put it:

> those who have been worried about the ways in which government action and social pressure limit liberty have been concerned about liberty in a wider sense than closing off alternatives or rendering them more costly. Let us call the other aspects of this wider sense of liberty, "autonomy,"—the control an individual has over his or her own evaluations and choices. If one is concerned with autonomy as well as freedom, narrowly conceived, then there does seem to be something paternalistic, not merely beneficent, in designing policies so as to take advantage of people's psychological foibles for their own benefit... The reason why nudges such as setting defaults seem ... to be paternalist, is that in addition to or apart from rational persuasion, they may "push" individuals to make one choice rather than another. Their freedom, in the sense of what alternatives can be chosen, is virtually unaffected, but when this "pushing" does not take the form of rational persuasion, their autonomy—the extent to which they have control over their own evaluations and deliberation—is diminished. Their actions reflect the tactics of the choice architect rather than exclusively their own evaluation of alternatives.   (2010, p. 128)

Hausman and Welch's point is that, while nudges may not interfere with our freedom of choice, they clearly interfere with our control over our deliberations. But this is just to say that they interfere with our autonomy. Further, nudges interfere with our autonomy precisely because they are *not* methods of rational persuasion. They 'take advantage' of our 'psychological foibles' to produce a desired result.

I do not think the science marketing methods I discussed in §4.2 are clear examples of nudges. But they are similar in that they are not methods of rational persuasion. Like nudges, they involve presenting information in ways designed to make it more likely that the audience will react in a desired way (i.e. form accurate views about global warming). We can also make sense of how they work in terms of 'choice architecture'. The idea is to shape the direction of inquiries (the choices inquirers make) and the materials inquirers have to hand (the sources of information they choose to consult) in such a way that these inquiries are liable to reach particular conclusions.

To illustrate these points, take inoculation theory. The idea behind inoculation theory is that you can inoculate someone against misinformation by leading them to think about things in a certain way. This is, effectively, to interfere with the choices people make as inquirers, such as their choices about which issues to think about and which questions to ask. Similarly, take framing. An issue like global warming can be framed in a way designed to secure maximum uptake. This means that, while inquirers may not be forced to adopt any particular conclusion, the way in which they think about the issue—the way that they frame it—is decided for them. Moreover, it is decided for them in such a way as to lead them towards a particular conclusion, that is, that global warming is real, and humans are responsible for it.

In summary, it looks like, if nudges are problematic because they infringe on our autonomy, the science marketing methods I have discussed are problematic for the same reason. My aim in the next two sections is to argue that, appearances notwithstanding, there are some significant differences between nudging and science marketing. While Thaler and Sunstein propose a systematic programme of nudging that may well infringe on our intellectual autonomy, it is less clear that the more targeted interventions proposed by science marketers infringe on intellectual autonomy. I will make my case by considering two papers that argue against nudging on the grounds that it infringes on our intellectual autonomy. I will argue that what these papers really show is that a systematic programme of nudging infringes on autonomy. Because science marketing is far more targeted, it need not do so.

4.4 Riley on Nudging and Epistemic Injustice

In his 2017 paper 'The Beneficent Nudge Programme and Epistemic Injustice', Evan Riley argues that the beneficent nudge programme is problematic on both ethical and epistemological grounds. In this section, I will outline his argument, before drawing out what it shows—and does not show—about intellectual autonomy. Here is Riley summing up his argument:

> [Thaler and Sunstein] do not simply hold that nudging is in some circumstances morally permitted and practically called for. Rather, they favor and would have us foster the broad adoption of the beneficent nudge, to be deployed as a general purpose tool for good, across the institutional milieu of contemporary social life. In a recent defense of implementing this—which I call the beneficent nudge program (BNP)—Sunstein has gone so far as to christen and defend the 'First Law of behaviorally informed regulation: *In the face of behavioral market failures, [beneficent] nudges are usually the best response, at least when there is no harm to others*'. (2017, p. 598)

Riley's target is not nudging per se but the 'beneficent nudge programme'. The beneficent nudge programme combines two elements. First, nudges should be 'deployed as a general purpose tool for good'. For Thaler and Sunstein, this means that nudges should be deployed to help people make the decisions they would make if they were better at figuring out what was in their long-term interest. Second, nudges should be employed widely (see Sunstein's First Law). This is what makes the beneficent nudge programme a *programme*. The beneficent nudge programme goes beyond the idea that nudges can sometimes be appropriate. The idea is that we should engage in a systematic programme of nudging.

We can now look at Riley's objection to the beneficent nudge programme in more detail. His point is that, while nudges do not exactly bypass the 'nudgee's' (the person who is nudged) critical faculties, they also do not engage them fully. When someone is nudged, they may be made aware of important new information. In our example with the doctor and patient, the patient is made aware of some relevant statistics about the outcome of the operation. They may also engage in rational inference. The patient might compare the survival statistics for the operation with what they know about their chances of survival given the disease they have. But the crucial point is that the patient's reflective critical capacities are not fully engaged because they are steered towards a particular conclusion by having the information presented to them in a way designed to invite that conclusion.

It may be that being nudged occasionally does little to reduce your capacity for critical reflection and deliberation. You will have plenty of other opportunities to fully engage your critical capacities. But being subjected to a systematic programme of nudging may well reduce this capacity. Being subjected to a systematic programme of nudging may well prevent you from developing the capacity to 'reason critically, energetically and otherwise well' (Riley 2017, p. 604). Riley thinks that this may even be a form of epistemic injustice:

denying or neglecting to provide people the support, opportunities, or means necessary to develop those capacities [e.g. the capacity to reason critically], or making it relatively more difficult to develop and exercise those capacities, where this lack could be supplied or ameliorated without duly weighty sacrifice or some other comparatively serious consideration, is unjust. In addition, the character of this general kind of wrong cannot be made fully explicit without reference to the epistemic nature of the victim. Thus, it counts as an epistemic injustice. Call it reflective incapacitational injustice.   (2017, p. 605)

For my purposes we do not need to worry too much about the ins and outs of what Riley means by a 'reflective incapacitational injustice'. All we need is the central claim that, while being subjected to the occasional nudge may not do much harm, being subjected to a systematic programme of nudging deprives you of the opportunity to develop the capacity to reason critically. Whether this is an epistemic injustice in the sense of Fricker (2007) doesn't affect this point.[3]

This completes my overview of Riley's argument. I now want to relate it more directly to science marketing and intellectual autonomy. But first a preliminary remark. Riley talks about the capacity to reason critically rather than intellectual autonomy. But clearly these things are connected. Roughly speaking, we can say that a capacity to reason critically is an essential part of being intellectually autonomous. I will leave it open just how much more there is to intellectual autonomy than this. On some views (e.g. the view defended in Zagzebski 2013) they seem to come to almost the same thing.

In §4.3, I suggested that, while the science marketing strategies I have discussed may not be nudges strictly speaking, they are broadly similar. The crucial question is whether attempts to use them run into the same problems Riley thinks the beneficent nudge programme runs into. Riley's discussion of the beneficent nudge programme points to a way in which the use of these science marketing strategies might be defended from the charge that they infringe on intellectual autonomy.

Riley distinguishes between interfering with someone's ability to fully engage their reflective critical capacities in a particular case and preventing them from developing the capacity to reason critically. In nudging someone, you may interfere with their ability to reason critically at that point in time, but this hardly prevents them from developing the capacity to reason critically. His point is that *over-use* of nudging as a strategy has this problematic result, not that every nudge

---

[3] I am inclined to think that it can but need not be an epistemic justice in Fricker's sense. For Fricker, an epistemic injustice is a wrong done to someone in their capacity as a knower (2007, p. 1). The idea of a 'reflective incapacitational injustice' fits nicely with this definition. If someone stops me developing the capacity for critical reflection and deliberation, they certainly wrong me in my capacity as a knower. But Riley does not discuss the sorts of identity prejudices and stereotypes that play a vital role in Fricker's discussions of testimonial and hermeneutical injustice. My suspicion is that the beneficent nudge programme leads to a form of epistemic injustice only when it is based on identity prejudices and stereotypes about the targets of the programme.

does. Put in terms of intellectual autonomy, Riley's point is that interfering with someone's deliberations in a particular case is not the same thing as infringing on their intellectual autonomy. Intellectual autonomy is a *capacity* and preventing someone from manifesting a capacity at one point in time is not the same thing as preventing them from *having* or *developing* the capacity (for a similar point, see Feinberg 1986, pp. 27–51).

If this is right, then it is at least possible to argue that the science marketing strategies I have discussed need not stop anyone from becoming intellectually autonomous. But is it more than possible? The first thing to say is that, while the proponent of the science marketing methods I have discussed is putting forward a programme, it is a more limited programme than Thaler and Sunstein's beneficent nudge programme. Thaler and Sunstein envisage using nudges in all areas where humans do a bad job of making choices and decisions that serve what many would regard as their interests. The idea behind the science marketing strategies is that they can nudge people who are otherwise unwilling to accept the science on global warming and similar issues to change their minds. To the extent that Riley's objection to the beneficent nudge programme is based on how extensive it is, it does not carry over to science marketing because it is less extensive and more targeted.

The second thing to say is that the aims of those who want to better market science differ from the aims of commercial marketers. This remark by Dan Kahan (a proponent of science marketing strategies) is suggestive:

> It would not be a gross simplification to say that science needs better marketing. Unlike commercial advertising, however, the goal of these techniques is not to induce public acceptance of any particular conclusion, but rather to create an environment for the public's open-minded, unbiased consideration of the best available scientific information.   (2010, p. 297)

It may be that Kahan overstates the point here. It certainly seems like *one* of the goals of science marketing is to induce public acceptance of certain conclusions (that global warming is real, that humans cause it, etc.). But the point still stands that this goal need not be in tension with the goal of creating a better epistemic environment—an environment where people can reason critically about issues like global warming based on the best available scientific information.

Ultimately, what the science marketer is trying to create is an environment in which the public can reach the conclusions that are supported by the evidence, which is that global warming is real and is caused by humans. So we can say something like this. Surprisingly, the best way to help people develop the capacity for reasoning critically about scientific issues is to infringe on their deliberations about particular issues. Infringing on their deliberations can, in certain cases, and if it is done properly, further rather than hinder the development of intellectual autonomy. I will leave this point for now, but I expand on it in §5.6.[4]

4.5 Meehan on Nudging and Epistemic Vices

I have just argued that, while Riley may succeed in showing that the beneficent nudge programme infringes on intellectual autonomy, his argument points to some reasons for thinking that science marketing need not do so. In her 2020 paper 'Epistemic Vices and Epistemic Nudging: A Solution?', Daniella Meehan makes a similar point to Riley but in a different way. It is therefore worth seeing whether her argument runs into the same issues as Riley's, at least when it is applied to science marketing. Where Riley argues that the beneficent nudge programme is problematic on the grounds that it leads to a distinctive sort of epistemic injustice, Meehan argues that the programme is, at best, minimally effective in combatting our existing epistemic vices and, at worst, leads to the creation of new epistemic vices. To provide some focus we can start with two of her core cases, one of which is fictitious and the other of which is drawn from the real world.

In the fictitious example, an epistemic agent, Harry, forms various false beliefs about politics and related matters because he makes extensive use of unreliable news sources and is unwilling to look at more reliable sources. Harry, in short, is not only closed-minded but closed-minded about his closed-mindedness. His vice is 'stealthy' in the sense that it 'blocks' its own detection (Cassam 2019). Meehan asks: what can we do about Harry? We could sit him down and lay out the reasons why he is being an irresponsible epistemic agent. Or we could try nudging. We could, for example, offer him a discounted subscription to a more reliable paper or leave unbiased news programmes on TV. These are nudges because, while they are interventions that are intended to produce a desired outcome, he is still free to decide for himself what he thinks about all these issues.

For her real-world example, Meehan looks at Michigan's use of the 'inconvenience model' for tackling falling rates of child vaccination. The basic idea behind the inconvenience model, as practised in Michigan and elsewhere, is to put barriers in place to securing an exemption from the requirement to vaccinate children before sending them to school, kindergarten, and so on. In Michigan, parents had to attend educational sessions about vaccines at local public health centres and to use

---

[4] Levy (2021, ch. 6) argues for similar conclusions about nudging and intellectual autonomy but on the basis of slightly different considerations. For Levy, nudging need not infringe on intellectual autonomy because, when you are nudged, the right sorts of considerations may guide you—the considerations that ordinarily make for rational belief. While I am inclined to agree with this (for more, see §5.6), I have focused more on the implications of nudging for our capacity to be guided by the right sorts of considerations.

an official state form to apply for exemptions. This had a demonstrable impact on exemption rates, which went down by 39 per cent state-wide and by 60 per cent in the Detroit area (Higgins 2016). This is an example of nudging because, while the intent behind these policies was to produce a particular outcome (i.e. a reduction in the exemption rate) parents were still free to not vaccinate their children if they so wished.

Meehan argues for two main claims, which she illustrates using these two cases. Her first claim is that, even if nudging is successful in tackling epistemic vices in the short term, it is ineffective in the long term. Here is her basic argument ('EN' refers to 'epistemic nudging', which is nudging that targets our activities as inquirers):

> When EN claims to have successfully mitigated a vice, like the examples presented earlier, what has really happened is that EN has merely *masked* the epistemic vice at hand. EN can only mask epistemic vices as the deep nature of vices remains present. EN does not change the vice in any way, just like the bubble wrap did not change the fragility of the vase, but only masks it, and when EN practices are not employed the vice is still present, just like how the fragility of the vase still remains when the bubble-wrap is removed.   (2020, p. 253)

In Meehan's first example, even if your attempts to get Harry to consult more reliable news sources are successful in the short term, they are merely masking his closed-mindedness, which will inevitably resurface in the future. The case is, like any fictitious case, under-described. But it is easy to see what Meehan has in mind here. It is a lot easier to change behaviour in the short term than it is to change underlying character traits. Because nudging seems more focused on changing behaviour in the short term than on changing underlying character traits, it is easy to see why it would merely mask underlying vices.

It is a little harder to evaluate Meehan's claim in her second example because it really depends on why the 'inconvenience model' was successful in Michigan. Did it change anyone's minds, or did it work purely because it made avoiding vaccination so inconvenient (Navin and Largent 2017)? Perhaps the right thing to say is that the onus is on Meehan's opponent to give some reasons for thinking that the model really did change minds. If it did not even change the minds of the parents who decided to vaccinate their children, then it is hard to see how it could have altered their intellectual character traits.

Let me say two things about Meehan's claim that nudging merely masks intellectual vices. First, Meehan is asking too much of nudging. It would be surprising if nudging were enough to change someone's underlying character traits. If we are to imagine that Harry is genuinely closed-minded, then your attempts to steer him in the direction of better media sources is not going to be enough to change this.

However, nudging does not need to change anyone's underlying character traits to do more than merely mask epistemic vices. As Quassim Cassam (2019, p. 7) emphasizes, it can be incredibly difficult to get started with tackling our epistemic vices. Harry's problem is not just that he is closed-minded. It is also that he is not well-positioned to recognize his closed-mindedness. One thing nudges could do for Harry is steer him towards recognizing this fact about himself. If Harry is suddenly in a situation where he is regularly encountering views that run contrary to his own, this may prompt some critical reflection. You would not necessarily expect him to change overnight. But you might well expect him to recognize that, in the past, he has not even considered alternative points of view. This is a step—though only a step—towards doing something about his closed-mindedness.

Second, and more importantly, my primary focus in this chapter is the problem of consequential false beliefs: how can we persuade global warming sceptics and other 'science denialists' to change their minds? Let us take Harry. Given his media consumption habits, Harry may well have some beliefs that we would want to change. Meehan is not (as far as I can see) disputing whether nudges can bring about a change of beliefs. If you arrange things so that Harry now gets his news from more reliable sources, he is going to change some of his beliefs. Meehan is disputing whether nudges can change Harry's basic character traits. If he was closed-minded before, he will be closed-minded now. If you had a way of changing basic character traits—a way of making the closed-minded more open-minded, for instance—this would be a great help with the problem of consequential false beliefs. But the fact that nudging cannot alter our basic character traits is no reason to think that nudges are not a practical and straight-forward way of solving the problem of consequential false beliefs.

We can now move on to Meehan's second claim, which is that nudging may even foster new epistemic vices, such as the vice of epistemic laziness. More generally, nudging may hinder our epistemic capacities. Meehan is getting at a similar idea to Riley, as can be seen from this passage:

> EN does not merely accidentally fail to engage the critical deliberative faculties of their targets, but purposefully seeks to bypass reflective deliberation entirely. Take the educational tool... where teachers nudged students towards effective inquiry by teaching incomplete theories to facilitate a better understanding of their complexities. As Riley would note, this form of EN seeks to bypass a genuine open reflective deliberation, meaning for one to nudge successfully in this case (and many others) one must at the time of the nudge, not invite, seek or start any critical reflection or deliberation.   (2020, p. 254)

Where Riley puts the point in the language of epistemic justice, Meehan puts the point in terms of epistemic vices. But they are getting at a similar idea. In not engaging the nudgee's critical deliberative faculties, the 'nudger' (the person doing the nudging) may well foster an unwillingness or even inability to inquire for oneself (epistemic laziness) in the nudgee. If this were to happen, the nudgee's capacity for critical reflection and deliberation would be diminished.

Like Riley, Meehan is pointing to the fact that nudging can interfere with our intellectual autonomy. But, as we saw in the previous section, Riley does not think that 'one-off' nudges need interfere with anyone's intellectual autonomy. His point is that subjecting someone to a systematic programme of nudging (the beneficent nudge programme) will interfere with their intellectual autonomy because it will make it harder for them to become intellectually autonomous. Meehan seems to be making the stronger claim that nudging itself interferes with intellectual autonomy. But there is nothing in her argument to support drawing this stronger conclusion, as opposed to the weaker conclusion that Riley draws. Just as it is hard to see how being subjected to 'one off' nudges could prevent someone from becoming intellectually autonomous, it is hard to see how 'one-off' nudges could create the vice of epistemic laziness. (It is also hard to see how the educational tool mentioned in the quoted passage could create any vice at all— isn't this simply good pedagogy?)

In conclusion, Meehan's argument at best supports the claim that we saw Riley makes, which is that over-use of nudging can prevent someone from becoming intellectually autonomous. Relating this back to science marketing, we are left with the same conclusion: we can argue that, because the science marketer is proposing a more limited programme than Thaler and Sunstein, it is less clear that what they are proposing infringes on anyone's intellectual autonomy. The point, again, is that there is a difference between interfering with someone ability to deliberate autonomously in a particular case and preventing them from developing the capacity for intellectually autonomous thought.

4.6 Tsai on Rational Persuasion and Paternalism

So far, I have looked at some reasons for thinking that nudges and science marketing strategies are problematic. I have argued that, while Riley and Meehan might succeed in showing that widespread and systematic use of nudging infringes on our intellectual autonomy, judicious use of science marketing strategies need not do so. In this section, I take a different approach and attack a central assumption that often underlies arguments against paternalism, whether epistemic or not, which is that there is a crucial difference between paternalism and *rational persuasion*.

It is tempting to think that there is a fundamental difference between rational persuasion and paternalism. Rationally persuading someone to do (or think) something involves offering reasons, evidence, and arguments, whereas paternalistic interference involves manipulation of some sort (Berofsky 1983; Groll 2012;

Shiffrin 2000). In the present context, the thought would be that there is a crucial difference between persuading someone to accept that human activity is the major cause of global warming through the provision of reasons, evidence, and arguments, and persuading them to accept this through science marketing. I am going to argue against this central assumption. Rational persuasion and certain forms of paternalistic interference have more in common than you may think.

My argument is based on George Tsai's 2014 paper 'Rational Persuasion as Paternalism'. Tsai also argues that rational persuasion and paternalism need not be as different as many think, but he draws the opposite conclusion to me. Where I conclude that paternalism is less problematic than many think, Tsai concludes that rational persuasion is more problematic than many think. I will start by looking at Tsai's argument before explaining why I take it to support my conclusion rather than his. Here is Tsai stating his basic claim:

> [I]t is possible to rationally persuade someone to do something, yet treat her paternalistically...Rational persuasion may express, and be guided by, the motive of distrust in the other's capacity to gather or weigh evidence, and may intrude on the other's deliberative activities in ways that conflict with respecting her agency.   (2014, p. 79)

Tsai's claim is that rationally persuading someone to do something like stop smoking can be problematic for the same reasons that paternalistically interfering with them to get them to stop (e.g. by hiding their cigarettes) is problematic. More specifically, Tsai holds that there are two problematic aspects of paternalistic interference. The first is that it is guided by what he calls a 'motive of distrust' in the rational capacities of another person (the person being interfered with). The second is that it involves interfering with the deliberations and decisions of another person in ways that disrespect their autonomy. Tsai's claim is that some cases of rational persuasion also have these two aspects, and so are problematic for the same reasons that paternalistic interference is.

For example, Claire is trying to decide whether to go to graduate school in philosophy or to law school. Her father, Peter, is trying to persuade her to go to law school. To this end, he bombards Claire with information about the multiple ways in which law school is the better choice. Peter is trying to persuade Claire to decide to go to law school through rational means; he is not extorting her, tricking her into going, or anything of the sort. But Tsai thinks his actions are problematic because they interfere with Claire's ability to make the decision for herself:

> when others offer us reasons to persuade us at the wrong time or in the wrong way, they make it harder for us to be able to engage more purely and directly with the reasons most centrally tied to the choice-worthiness of our options. When our deliberations are *distorted* in this way, this potentially alters the self-determining

and self-expressive aspects of our decision...the point is that even the rational pressure of Peter's *reason-giving* (as distinguished from the rational pressure of the *reasons* themselves) might potentially alter the nature of Claire's deliberations in a way that results in a sense of loss for Claire...Insofar as the timing of Peter's attempt at rational persuasion precludes Claire from having the purer, more direct engagement with the reasons most centrally relevant to her deliberative situation, this limits her exercise of epistemic agency.   (2014, pp. 95–6)

Tsai is right that Peter's behaviour is problematic, but wrong in thinking that this example generalizes in the way needed to establish the conclusions which he draws from it. Peter is interfering with Claire's ability to deliberate and decide for herself because he has no trust in her ability to make this decision in a competent manner. Given this, whether his interference takes the form of rational persuasion (the offering of reasons, evidence, etc.) or something more manipulative makes little difference. But it is hard to see why the example should generalize. Rational persuasion, like paternalistic forms of interference, may be accompanied by a lack of trust and this lack of trust may often be disrespectful. But it need not always be disrespectful (our capacity to deliberate and decide can be bad) and rational persuasion need not always be accompanied by a lack of trust.

First, part of the reason Peter's distrust in Claire's ability to make this decision is disrespectful is that the decision in question (whether to go to law school or graduate school in philosophy) concerns Claire's life and what she wants to do with it. Peter's distrust in her ability to make this decision for herself reflects a lack of respect for Claire's deliberative capacities and for her ability to make life choices. It is therefore unclear whether Tsai's point will generalize to cases where the decision has nothing to do with the life choices of the deliberator.

Importantly, this point stands whether these cases involve rational persuasion or something more paternalistic. What Tsai has really shown is that the extent to which a lack of trust in someone's ability to deliberate about some matter is disrespectful depends more on what they are deliberating about than on the way in which you go about trying to change their choices and alter the course of their deliberations.

Second, there are many ways of interfering with someone's deliberations and decisions. Tsai's fundamental objection to Peter's behaviour seems to be that he has made it harder for Claire to engage with the reasons for and against her two choices. But this is a consequence of the way in which Peter has chosen to interfere with Claire's deliberations, not of him interfering at all. Peter could have 'stated his case' in a way that did not interfere with Claire's ability to engage with the reasons. He could have sat her down and told her why he thought law school was the right choice but made it clear that it was her decision to make, and he was just there to provide her with information, not to decide for her. Alternatively, he could have ensured that she got all the information she needed through less overt means, like leaving marketing brochures for each course lying around.

Again, it is important to highlight that what Tsai has really shown is that what matters is whether you make it harder or easier for someone to engage with the reasons for and against the various choices they are weighing up. While rational persuasion may usually do a better job of facilitating this sort of engagement, there is no reason why more paternalistic forms of interference cannot facilitate it too.

Up to this point I have been making the case that Tsai has not shown that his example generalizes to a wider class of cases of rational persuasion. Along the way I have suggested that the reasons why it does not generalize also help to see why more paternalistic forms of interference need not be as problematic as many seem to think. But let me now make the case that paternalism need not be problematic more explicitly. I will take the science marketing strategies I have discussed (framing, choice of spokesperson, prebunking) and show that they can avoid the problems Tsai identifies in his example. If Tsai has identified what makes rational persuasion or paternalism problematic, then the fact that science marketing need not have these problematic features is a good reason to think it need not be problematic.

First, our deliberations about what to believe when it comes to global warming do not directly concern our life choices, though they do so indirectly, in that what we believe about global warming may inform our life choices. Interfering with someone's decisions about what to think about global warming therefore need not reflect any disrespect for their ability to make life choices. The point is not to persuade anyone to choose differently. The point is to give people the information they need to make informed choices—and to get around the barriers that exist to them taking on board the information that is already available.

Second, the science communicator who uses these strategies is not aiming to interfere with people's abilities to engage with the scientific evidence about global warming. On the contrary: their aim is to facilitate engagement with this evidence, by drawing attention to what the evidence is and neutering ideological biases and blind spots that get in the way of engaging with it. As I have suggested in a few places, the aim of these strategies is different from the aim of commercial marketing strategies. The goal is to create an environment where people can consider the available scientific evidence (recall the Kahan quote discussed in §4.4).

Third, and finally, while the empirical literature I have canvassed in this chapter does engender distrust in the capacity of laypersons to form views about complex issues like global warming, this distrust is based on the evidence provided by the empirical literature. Moreover, the distrust is in our capacity to form views about these issues *given certain widespread features of human psychology and in the complicated socio-epistemic environment in which we find ourselves.* This sort of distrust need not involve any lack of respect for laypersons' epistemic agency. It is a simple consequence of the realization that we are imperfect epistemic agents in an imperfect world. To recognize this is not disrespectful and to refuse to recognize it is not particularly respectful.

This links with the more general theme of this book, which is that the various respects in which we fall short of being perfect epistemic agents matter. If an agent has an ability and can manifest it in the environment in which they find themselves, then not recognizing that they have the ability is clearly disrespectful, absent good reasons for not recognizing that they have the ability. On the other hand, if most agents lack an ability (or are unable to manifest it in the environment in which they find themselves) then it is unclear why not trusting them to exercise the ability is disrespectful. This is especially so when the lack of trust is simply based on the realization that most agents lack the relevant ability. Of course, you can manifest distrust in disrespectful ways, and the reasons you have for distrust can be bad ones. But why think distrust is itself disrespectful, or that it is impossible to manifest it in ways that are respectful?

I conclude that, while Tsai is right in arguing that rational persuasion is not necessarily that different from paternalistic interference, he draws the wrong moral. Tsai concludes that rational persuasion is often as problematic as paternalistic interference. I have argued that, on the contrary, in some cases neither rational persuasion nor paternalistic interference is problematic.

4.7 Intellectual Autonomy

Let me sum up. My aim in this chapter was to address what I called the problem of consequential false beliefs. But, as I said at the outset, we need to distinguish between two questions. The first, which we can only answer by looking at relevant empirical literature, is which strategies for persuading people to change their minds are effective. The second, which calls for philosophical reflection, is whether we are permitted to use the methods we think are likely to be most effective.

In this chapter, I have addressed both questions. In answer to the first, I surveyed literature on science communication and the psychology of persuasion. As a result of this survey, I identified three strategies that are likely to be effective in changing people's minds about global warming and similar issues at the intersection of science and politics. These strategies look at how science might be better marketed, and as such can aptly be described as science marketing strategies. My answer to the first question goes some way towards completing the first task I said at the end of the previous chapter that Anderson left unfinished. It yields a set of suggestions for constructing a better epistemic environment that are evidence-based and so fit within a non-ideal approach to institutional epistemology.

My answer to the second question was a little more involved. I tried to show that science marketing need not infringe on anyone's intellectual autonomy. I also questioned the common assumption that there is a significant difference between methods of rational persuasion (which are permissible) and more paternalistic forms of interference (which many see as impermissible). My answer to the second question also goes some way towards the second task that Anderson left unfinished: it addresses the concern that non-ideal institutional epistemology proposes a problematic form of interference with our inquiries.

There remain two big open questions. The first, which I am not going to address here or in this book, has to do with the way in which Anderson initially frames the problem of identifying experts and I have framed the problem of consequential false beliefs. They were framed as problems within democratic political theory: given that you need to have broad public support for science-based public policy, what can you do when certain sections of the public reject the very science on which public policy might be based? I have argued that a plausible set of strategies for dealing with this problem (i.e. the science marketing strategies I have discussed) do not infringe on intellectual autonomy. But there are other reasons you might have for thinking that they are in one way or other democrat-ically illegitimate. So I cannot claim to have fully resolved the political problem which Anderson highlights.

The second question, which I address in the next chapter, concerns intellectual autonomy itself. I have taken it for granted that intellectual autonomy is important and valuable. My task has been to show that science marketing need not infringe on it. But you might think that there is something problematic about intellectual autonomy itself. It is a familiar point that certain conceptions of intellectual autonomy are problematic because they conceive of it in such a way that it is simply unattainable, or in such a way that it isn't clear why we should value it (Carter 2020). However, those who criticize conceptions of intellectual autonomy on these grounds typically propose a different way of conceiving of it—a way on which it is both attainable and valuable. In the next chapter, I criticize these 'modest' conceptions of intellectual autonomy. As it turns out, we should often not strive to be intellectually autonomous even on a more modest concep-tion of what being intellectually autonomous would involve.

5

Intellectual Autonomy

In the previous chapter, I set the issue of the value and importance of intellectual autonomy aside. In this chapter, I address it directly. My claim is that intellectual autonomy is not just an epistemic ideal or goal that many of us frequently and predictably fall short of. More importantly, it is an epistemic goal that often frustrates our other epistemic goals in the sense that, in striving to attain it, we run the risk of not achieving these other, often more valuable, goals. I conclude that, in the absence of a reason for thinking that intellectual autonomy is more important than our other epistemic goals, we often do better not to strive for it.

This has implications for the debate about epistemic paternalism. In this debate, intellectual autonomy functions as an ideal, in that it puts strict limits on the extent to and ways in which we can interfere with the inquiries of others. But, if we sometimes do better not to try and preserve intellectual autonomy, then, as I will argue, objections against epistemic paternalism lose much of their force. This chapter therefore buttresses my argument in the previous chapter in defence of science marketing. Even if you are not convinced that the 'science marketer' can avoid infringing on intellectual autonomy, my argument in this chapter shows that it is not necessarily a problem if science marketing infringes on intellectual autonomy.

Let me be clear from the outset that my target in this chapter is not what I call 'radical' conceptions of intellectual autonomy, on which it requires near complete epistemic self-reliance. Many would agree that 'radical intellectual autonomy' is not a goal worth striving for. My target is what I call 'modest' conceptions of intellectual autonomy. On modest conceptions, intellectual autonomy requires (as Roberts and Wood 2007 put it) 'wise' epistemic dependence. The point of modest conceptions is supposed to be that they think of intellectual autonomy in such a way that it is both an attainable and valuable goal. My aim is to cast doubt on whether even modest conceptions of intellectual autonomy can succeed in doing what they are intended to do.

This chapter illustrates what I called in Chapter 1 the third key aspect or face of non-ideal epistemology, which is a general objection to ideal epistemology. The objection is that the norms of inquiry or epistemic ideals proposed by the ideal epistemologist are often not just unattainable but, more importantly, serve to frustrate our other epistemic goals. In this chapter, I argue that intellectual autonomy is like this. It is often the case that, in striving for it, we miss out on

Non-Ideal Epistemology. Robin McKenna, Oxford University Press. © Robin McKenna 2023.
DOI: 10.1093/oso/9780192888822.003.0005

true beliefs and bits of knowledge we would otherwise have had. This does not mean that we should reject the ideal of intellectual autonomy wholesale. But it does mean we should not always try to be intellectually autonomous. One of my goals is to identify some places where we do better to not strive for intellectual autonomy.

Here is the plan. I start by saying what intellectual autonomy is and distinguishing between radical and modest conceptions of it (§5.1). In the following three sections, I set radical conceptions aside and argue against two modest conceptions of intellectual autonomy (§§5.2–5.4). I then relate my argument in this chapter to the argument of Chapter 4 and explain what the upshots are for epistemic paternalism (§5.5). Finally, nothing I say in this chapter conflicts with the view that, in a (more) epistemically ideal world, we would all be intellectually autonomous. I finish the chapter by considering if there are ways of getting closer to such a world (§5.6). I suggest that, paradoxically, epistemic paternalism may offer the beginnings of a way to getting closer to it.

5.1 What Is Intellectual Autonomy?

We can start with what intellectual autonomy is. Put roughly, the intellectually autonomous individual is able and willing to think for themselves. They have a mind of their own. But what does thinking for yourself involve? In this section, I briefly discuss and reject 'radical' views of intellectual autonomy, before going on to consider two 'modest' views in more detail. These views are modest in that they are designed to avoid the worries many have raised about more radical conceptions of intellectual autonomy. They therefore need to be taken seriously even by those who are inclined to reject these more radical conceptions.

Radical views of intellectual autonomy involve a combination of the following ideas:

- Intellectually autonomous agents rely on themselves to get information about the world (Hume 2007; Fricker 2006).
- Intellectually autonomous agents 'have the courage to use their own intelligence' (Kant 1959).
- Intellectually autonomous agents do not allow their beliefs to be shaped by what others think (Emerson 1841).

If we combine these ideas, we get a very radical view of intellectual autonomy indeed. On this view, the intellectually autonomous individual finds things out for themselves, figures out how to do things for themselves, and pays no attention to what anyone else says or thinks. As Elizabeth Fricker puts it:

This ideal type [the intellectually autonomous agent] relies on no one else for any of her knowledge. Thus she takes no one else's word for anything, but accepts only what she has found out for herself, relying only on her own cognitive faculties and investigative and inferential powers.   (2006, p. 225)

There would be *something* admirable and impressive about someone who was like this. But, if any normal human were to try and emulate this 'ideal type', they would deprive themselves of all sorts of epistemic goods. I can get some information myself, but I will get a lot more if I make proper use of information acquired by others. I should have the courage to use my own intelligence, and a certain independence of mind and thought is a good thing. But I also should have the humility to recognize my limitations. I know some things, but there is also a lot I do not know. I am good at some things, but I am bad at others. I will often do better epistemically if I defer to those who know more than if I push ahead on my own.

You do not need to have much sympathy with non-ideal epistemology to think it is a problem that 'radical intellectual autonomy' is unattainable for normal humans. In the literature, radical conceptions of intellectual autonomy are generally rejected precisely on the grounds of their unattainability (Carter 2020; Roberts and Wood 2007; Zagzebski 2013). But, while radical views go too far in one direction, you might think you can go too far in the other direction too. That is, you might think that you can be too epistemically dependent on others. What I call 'modest' views of intellectual autonomy try to find the middle-ground between the extreme self-reliance recommended by radical views and the opposite extreme of total epistemic dependence. In the rest of this section, I introduce two modest views of intellectual autonomy, before considering their merits in the next three sections.

The first view is defended by J. Adam Carter in his 2020 paper 'Intellectual Autonomy, Epistemic Dependence and Cognitive Enhancement'.[1] Carter's view is that intellectual autonomy requires a capacity for intellectual self-direction:

the virtuously autonomous agent actually must rely on others, and outsource cognitive tasks as a means to gaining knowledge and other epistemic goods, *up until the point that doing so would be at the expense of her own capacity for self-direction.*   (2020, p. 4)

---

[1] This paper doesn't represent Carter's most recent thinking on the issue (see Carter 2022). In his more recent work, Carter argues for an 'intellectual autonomy condition' on knowledge. The basic idea is that you only have knowledge if your belief is formed in an intellectually autonomous way. On this view, the goal of intellectual autonomy cannot come into conflict with the goal of extending knowledge in the way in which I argue that it can in §5.2. But, if we accept Carter's intellectual autonomy condition, my argument may establish an even more troubling conclusion: we have a lot less knowledge than we ordinarily think. (There are connections here with the argument of Chapter 8.) Setting this aside, even if we accept Carter's intellectual autonomy condition, and there is a way of avoiding these troubling conclusions, the goal of intellectual autonomy can come into conflict with other epistemic goals, such as the goal of believing truths and avoiding falsehoods. Whether I come by a true belief in an intellectually autonomous manner or not, I still come by a true belief.

Carter's thought is that the intellectually autonomous individual keeps control over the direction and shape of their inquiries. But this control can take the form of deferring to others for information and outsourcing certain tasks. The underlying idea is that there is a difference between *deciding* to outsource cognitive work to someone else and having someone *take over* your cognitive activities. The former is, properly understood, entirely consistent with being intellectually autonomous. The latter is inconsistent with intellectual autonomy. Importantly, it should be possible for most of us to exercise control over when we outsource cognitive work. Carter's conception of intellectual autonomy is modest in that he conceives of it in such a way that most of us can realistically aspire to it.

Carter recognizes that there is a tension between intellectual autonomy and other epistemic goods, such as knowledge and true belief. For instance, he emphasizes that relying on others and outsourcing cognitive tasks (e.g. using a smartphone) are ways of extending our body of knowledge. He argues that intellectual autonomy needs to be understood in such a way that you can sometimes extend your knowledge in these ways without sacrificing your intellectual autonomy. But he also argues that there are limits:

> enhancement via intelligence augmentation, as when outsourcing cognitive tasks to smartphones and other gadgets, subjects us to constant framing effects which often go unnoticed. While such gadgets obviously aid us in acquiring knowledge quickly and seamlessly, they—as this line of argument contends—undermine our intellectual self-direction by (and in a manner that typically goes undetected) diminishing the contribution that our own biological cognitive faculties make towards the shape our inquiries take.   (2020, p. 11)

When you type a search term into Google, the auto-complete function suggests what you might be looking for. In Carter's terms, the auto-complete function reduces your control over the direction of your inquiry. This may impact on the eventual result of your inquiry (e.g. if it changes the results of your search). So, in this example, a form of cognitive outsourcing reduces your control over the direction of your inquiry and so infringes on your intellectual autonomy. But, clearly, using Google, even with the auto-complete function, is an effective way of extending your body of knowledge. The goal of intellectual autonomy can conflict with our other epistemic goals. The intellectually autonomous individual may sometimes 'lose knowledge' they would have had if they had sacrificed their intellectual autonomy.

I return to this tension in §5.2, but I now want to turn to a second modest view of intellectual autonomy. Carter's discussion, while illuminating, does not really tell us what a self-directed inquiry is like. What does someone who directs their own inquiries do? For an answer to this question, we can turn to the view of intellectual autonomy defended by Robert C. Roberts and W. Jay Wood in their 2007 book *Intellectual Virtues: An Essay in Regulative Epistemology*.

According to Roberts and Wood, intellectual autonomy is a 'disposition of balance between hetero-regulation and auto-regulation in intellectual practice' (2007, p. 1). It is not complete independence from others, but a 'wise dependence'. The intellectually autonomous individual draws on the knowledge of others when appropriate, but they also stand their ground against pressures to conform.

What does 'wise dependence' look like in practice? Roberts and Wood explain how the intellectually autonomous individual conducts themselves when:

i. Relying on others
ii. Dealing with criticism
iii. Deferring to experts

As far as (i) is concerned, the intellectually autonomous individual exhibits a 'wisdom about knowledge': they know what they know, and they know what they do not know. Further, when it comes to what they do not know, they know where to get the information they need, and what to do with it when they get it. As far as (ii) is concerned, the intellectually autonomous individual makes judicious use of criticism. They do not reject criticism when it is warranted. But they also do not cave in the face of it. Finally, as far as (iii) is concerned, the intellectually autonomous individual defers when appropriate, but they also recognize the limitations of expertise. Expertise is restricted to domains and even experts are fallible.

Roberts and Wood are engaged in what they call 'regulative epistemology'. In contrast to the 'analytic epistemologist', who aims to produce *theories* of epistemic goods like knowledge (or intellectual autonomy), the regulative epistemologist aims to sketch a *picture* of what the intellectually virtuous agent is like. While the extent to which Roberts and Wood really depart from analytic epistemology is debatable (analytic epistemologists can paint pictures too), the aim of their regulative epistemology is to provide guidance that real inquirers can use. It is therefore central to their picture of intellectual autonomy that it is an ideal that we not only should strive for but, under the right circumstances, can achieve.

In summary, modest views of intellectual autonomy conceive of intellectual autonomy as an attainable epistemic goal for which we should strive. They therefore sharply differ from radical views, which conceive of intellectual autonomy in such a way that it is hard to see how we could ever attain it, or why we would want to. In the next three sections, I argue against these modest views. But before getting to that a quick word on why I focus on these views, given that there are other ways of thinking of intellectual autonomy available, including ways that seem consonant with the non-ideal epistemologist's emphasis on social situatedness (e.g. Daukas 2019; Elzinga 2019; Tanesini 2022).

I set these alternative views of intellectual autonomy aside in part because they differ from the way in which epistemologists tend to think of intellectual autonomy. For instance, Benjamin Elzinga's account is noteworthy because he defends a relational conception of intellectual autonomy on which our social relations to others are the grounds of our autonomy rather than potential constraints on it. This contrasts with the way in which Carter or Roberts and Wood understand intellectual autonomy, on which we need to 'carve out' space for an individualistic conception of intellectual autonomy given the recognition that we are socially situated. As such, accounts like Elzinga's fit nicely with what in Chapter 1 I called the second key aspect or 'face' of non-ideal epistemology, which is a view of inquirers and their epistemic responsibilities as deeply socially situated. In the next two chapters (Chapters 6 and 7) I move from the institutional face of non-ideal epistemology to this second face. In doing so I take up the question of intellectual virtue (and vice) on the non-ideal epistemologist's picture. I do not focus on intellectual autonomy in these chapters, but I sketch a picture on which our responsibilities as inquirers are grounded in our social relations.

5.2 Against Carter on Intellectual Autonomy

In the next three sections I argue against modest views of intellectual autonomy like Carter's or Roberts and Wood's. I will argue that, even on these modest views, intellectual autonomy is an epistemic goal that many of us will frequently and predictably fall short of. Moreover—and more importantly—in trying to attain it we run the risk of not attaining our other epistemic goals, such as the goal of extending our body of knowledge or true beliefs. The unattainability of intellectual autonomy is not merely disappointing. Striving to attain intellectual autonomy is often actively harmful to our overall epistemic situation.

Let me start with Carter's view of intellectual autonomy. Recall that Carter thinks the intellectually autonomous individual controls the direction and shape of their inquiries. Now, to *be* intellectually autonomous it is not required that you always *exert* this sort of control. As I emphasized in Chapter 4, intellectual autonomy is a capacity. You can have a capacity without displaying or manifesting it in all situations. Carter is particularly concerned with the ways in which technology can prevent us from exerting the desired form of control. To use our earlier example, when you use Google's auto-complete function, your control over the direction of your inquiry is reduced. As a result, your inquiry is no longer intellectually autonomous:

> the salient explanation for why . . . certain inquiries take the shapes they do, and culminate in the beliefs that they do (correct or otherwise), is technological design rather than the individual's own preferences. For those who outsource certain kinds of inquiries entirely to menu-driven search apps (e.g., Yelp reviews,

Spotify music suggestions, etc.) the beliefs which serve as the termination of these inquiries might be very different if the menu choices were different... This is not to say that individual preferences play no explanatory role in such inquiries at all... Rather, the point is that in some of these cases, the subject is led to claim a mistaken level of ownership, oblivious to the framing effect and its influence in her belief formation. (2020, p. 19)

Carter's strategy is to argue that, while some new technologies compromise our ability to exert control over our inquiries, we still have opportunities to exert control and so we can be viewed as intellectually autonomous, at least to a significant degree. But it is worth asking whether anything might compromise our ability to exert control in a more fundamental way. It seems to me that something might. In their work on nudge theory (discussed in Chapter 4) Thaler and Sunstein highlight the importance of 'choice architecture'. Let me briefly recap what Thaler and Sunstein say about nudges and choice architecture before explaining why choice architecture poses a more fundamental challenge to intellectual autonomy (as conceived of by Carter) than new technology.

A nudge is 'any aspect of the choice architecture that alters behaviour in a predictable way without forbidding any options' (Thaler and Sunstein 2008, p. 6). Examples of nudges include:

- Default rules (e.g. automatic enrolment of new employees in company pension schemes to encourage them to save more).
- Product placement (e.g. making healthy foods more visible to encourage healthy eating).
- Warnings (e.g. graphic pictures on cigarette packets to discourage smoking).
- Reminders (e.g. text messages to remind you that you have an upcoming medical appointment).
- Provision of valuable information (e.g. providing people with detailed energy bills, providing credit card customers with full data about how much the card is costing them, or providing nutritional information for food).

Nudges often work by changing the information we have at our disposal and influencing the directions of our inquiries. For example, disclosing information about the cost of your credit card prompts you to consider whether it is worth having. While debates about nudges typically focus on various ethical and political issues (see Chapter 4), here I want to focus on a claim Thaler and Sunstein make in defending the use of nudges. This claim is that choice architecture is inevitable and ubiquitous. As they put it:

choice architecture is inevitable. Human beings (or dogs or cats or horses) cannot wish it away. Any store has a design; some products are seen first, and others are not. Any menu places options at various locations. Television stations are placed on different positions on the dial, and strikingly, position matters, even when the costs of switching are vanishingly low; people tend to choose the station at the lower position. A website has a design, which will affect what and whether people will choose.   (Thaler, Sunstein, and Balz 2013, p. 11)

When presenting information or a product, designing technology (even simple technology, like a remote control), or designing a physical or virtual space (like a website), decisions must be made about the architecture—how to lay things out. These decisions are going to have real and often predictable impacts on the choices that people who use these things make. For example, choosing to place one product in a prominent place on the shelf and another on a less prominent place is likely to lead to an increase in sales of the first product and a decrease in sales of the second.

If Sunstein and Thaler are right about the inevitability and ubiquity of choice architecture—and I think they are—then 'choice architects' exert a lot of control over us. One form this control takes is over the direction of our inquiries. When a company decides (or is told) to provide their customers with some information, such as information about their energy use or credit card bill, this influences the direction their inquiries will take. For example, they might look to see if they could reduce their bill if they switched to another energy supplier. Choice architects can therefore exert a lot of control over the direction of our inquiries. Further, they can do so without us being aware of it because we do not always think about why someone is giving us information, still less why they are giving us these bits of information rather than other bits of information.

If we take what Sunstein and Thaler say about choice architecture and apply it to Carter's account of intellectual autonomy, we get the result that our intellectual autonomy is infringed on whenever our inquiries are influenced by choice architecture. Given the inevitability and ubiquity of choice architecture, it turns out that our intellectual autonomy is infringed on a lot of the time—often without us realizing. Carter is therefore forced to accept that our intellectual autonomy is not just compromised when we make use of certain new technologies or avail ourselves of novel forms of cognitive enhancement. It is compromised whenever we enter physical or virtual spaces that have been designed by choice architects. Of course, there will be some spaces that have not been designed by choice architects (a park? a forest?), so there will be some spaces where we can exercise our intellectual autonomy. But the point is that our ability to exercise it will be severely restricted.

You might respond on Carter's behalf that, even if the ubiquity of choice architecture prevents us from manifesting our capacity to exert control over our inquiries, we—or at least most of us—still retain this capacity and so remain intellectually autonomous. But this conflicts with the way in which Carter himself seems to understand the challenge to intellectual autonomy. Carter does think that some new technologies pose a threat to intellectual autonomy. His strategy is to argue that, in general, new technologies do not prevent us from exerting the required degree of control over many of our inquiries. This suggests that he thinks that, if we were unable to exert the required degree of control in a wide range of situations, then our intellectual autonomy would be markedly diminished. But I have just argued that, in view of the ubiquity of choice architecture, we are unable to exert the required degree of control in a wide range of situations. It follows that, by Carter's own lights, our intellectual autonomy is markedly diminished.

So far, I have argued that, even on Carter's modest view, intellectual autonomy is an epistemic goal that we frequently and predictably fall short of, often without even realizing that we are falling short of it. But I do not just want to show that, even on modest views, intellectual autonomy is often unattainable. My complaint is that intellectual autonomy is an epistemic goal that often frustrates our other epistemic goals. In striving to be intellectually autonomous, we often worsen our epistemic situation. Consequently, it is often the case that we should not try to be intellectually autonomous in the first place. But why does this follow from what I have said?

To see why it follows, we need to go back to Carter's discussion of new technologies. Carter wants to avoid the conclusion that using new technologies is incompatible with intellectual autonomy because, if this were the case, the natural conclusion would be that intellectual autonomy is, to put it bluntly, over-rated. As he emphasizes, new technologies have the potential to massively extend our body of knowledge. Why would we want to sacrifice all this for the sake of intellectual autonomy? Carter's aim is to show that, while we might need to sacrifice some gains in knowledge for the sake of intellectual autonomy, the sacrifices are not so great that they are more trouble than they are worth.

This suggests that, if the sacrifices were far greater than he thinks, they would be more trouble than they are worth. If it turned out that the only way to salvage intellectual autonomy were to sacrifice whole swathes of knowledge, the calculations would turn out differently. But this is precisely what the ubiquity of choice architecture shows. If the very fact that we inhabit physical and virtual spaces that are designed by choice architects undermines our intellectual autonomy, then, even if you were able to avoid such spaces, doing so would mean sacrificing all sorts of epistemic goods. Recall some of the examples of nudges cited earlier:

- Graphic pictures on cigarette packets to discourage smoking.
- Text messages to remind you of an upcoming medical appointment.
- Provision of information about energy bills, credit card costs, and nutrition.

Even if we were able to avoid being influenced by these sorts of nudges—or at least to minimize their influence—why would we want to? Why value intellectual autonomy more than the good of having valuable information about the dangers of smoking, an upcoming medical appointment, energy costs, or how healthy food is?

More generally, why value intellectual autonomy more than the good of having true beliefs or knowledge about things that matter to us? It is not clear that we have any reason to think intellectual autonomy is more important than having true beliefs or knowledge about things that matter to us. The fact we might have to sacrifice having true beliefs or knowledge for intellectual autonomy is therefore surely a good reason to not try and be as intellectually autonomous as we can. I conclude that intellectual autonomy, at least on Carter's conception of it, is not only hard to attain but also an epistemic goal which, in striving to attain, we run the risk of not attaining our other epistemic goals. This does not mean we should never strive for intellectual autonomy. But, absent a reason to think that intellectual autonomy is more important than these other epistemic goals, it does mean that we should often not strive to attain it.

In the next two sections I turn to Roberts and Wood's view. I argue that similar criticisms to those I have levelled against Carter's view of intellectual autonomy also apply to Roberts and Wood's view. My argument is based on the empirical literature on motivated reasoning, which I briefly discussed in Chapters 3 and 4. In the next section, I provide a more in-depth overview of this literature, before using it as the basis of my argument in the following section.

5.3 More on Motivated Reasoning

The literature on motivated reasoning looks at the influence of our background beliefs, desires, motivations, and values on our information processing. Manifestations of motivated reasoning can be distinguished in terms of the motivations they serve (Molden and Higgins 2012). A large body of work documents the ways in which our desire to maintain a positive self-conception leads us to give more credence to information that confirms our perception of ourselves as kind, competent, and healthy than to information that challenges these perceptions. For example, in a classic 1987 paper, Ziva Kunda reports a study in which subjects read an article stating that caffeine consumption causes serious health problems in women but not in men. Women who were heavy caffeine consumers found the article less convincing than women who were light caffeine consumers. The desire of these women to maintain a perception of themselves as healthy drove them to downplay the evidence against their self-perception. (For similar results, see Bradley 1978; Ditto et al. 1998; Kunda 1990.)

Another large body of work focuses on the impact our political ideologies have on our thinking about political issues and scientific issues that have become politically contentious. Call this *politically* motivated reasoning. For example, in their classic 1979 paper, Charles Lord, Lee Ross, and Mark Lepper report a study in which subjects were given the same set of arguments for and against capital punishment. Their assessments of the strength of these arguments correlated with their existing views about the rights and wrongs of capital punishment. Put simply, subjects who were predisposed to object to capital punishment found arguments against capital punishments more convincing while those who were predisposed to accept it found arguments for it more convincing. Whether for or against, their desire to confirm their existing views about capital punishment led them to downplay the strength of arguments against their views and overplay the strength of arguments for them. (For similar results, see Bullock, Gerber, and Hill 2015; Kahan, Jenkins-Smith, and Braman 2011; Jost, Hennes, and Lavine 2013; Redlawsk 2002.)

Three aspects of motivated reasoning are worth emphasizing. First, many of us engage in motivated reasoning, at least some of the time. Now, you might expect that, the 'smarter' you are, the less likely you will be to engage in motivated reasoning. But this would be a mistake. The empirical evidence suggests that it is, if anything, *more* prevalent in 'smarter' subjects. For example, Charles Taber and Milton Lodge (2006) found that the more politically knowledgeable a subject was, the more likely they were to conform their assessment of arguments to their political beliefs. Taber and Lodge suggest this is because politically knowledgeable subjects have more information at their disposal. The more information you have at your disposal, the better you are at finding flaws in arguments with conclusions you don't like, and at seeking out information that confirms rather than challenges your existing views (Brown 1986; Burnstein and Vinokur 1977; Kahan 2013).

Second, as Taber and Lodge also emphasize, motivated reasoning leads us to put the work into uncovering problems with hypotheses and views we would like to reject, but it also leads us to put little to no work into uncovering problems with hypotheses and views we would like to accept. The result is that what we would like to be true (and what we would like to not be true) impacts on what we think is true by impacting on the shape of our inquiries—which questions we ask, and which questions we do not. The 'motivated reasoning paradigm' does *not* say that we can 'believe at will'. Rather, it suggests that biases distort the shape of our inquiries, so that we are highly motivated to look for evidence that supports our existing views but reluctant to look for evidence that cast doubts on them.

Third, one of the key results in the literature on politically motivated reasoning is that our political ideologies impact on our evaluations of who the experts *are*. Dan Kahan, Hank Jenkins-Smith, and Donald Braman (2011) present studies which show that subjects' assessments of the level of expertise of (fictional) scientists correlate with how well their positions on topics such as global warming and nuclear waste disposal conform to the subjects' political convictions. Put simply, we regard those who take positions we disagree with as having *less expertise* than those who take positions we agree with. More generally, there is evidence that we assess expertise in a domain in terms of fit with our beliefs in that domain (Boorman et al. 2013; Faraji-Rad, Samuelsen, and Warlop 2015; Schilbach et al. 2013). Strikingly, there is even evidence that this tendency extends to things that have nothing to do with politics, such as ability to categorize geometric shapes, or to fix a car (Marks et al. 2019).

This completes my overview of the literature on motivated reasoning. I now turn to the implications for Roberts and Wood's account of intellectual autonomy.

5.4 Against Roberts and Wood on Intellectual Autonomy

Roberts and Wood's intellectually autonomous individual is good at making decisions about who to rely on for information and which experts to listen to. They also deal well with criticism of their views. But what does the literature on motivated reasoning tell us about how good we are at these things?

The answer, in short, is that it tells us that we are not good at them. When making decisions about who to rely on for information, we often look for what we want to hear, and ignore what we don't (recall Kunda's 1987 study about caffeine consumption). We also look for a 'fit' between our political beliefs and values and the political beliefs and values of our potential informants. This applies even when we are making decisions about which experts to rely on (recall Kahan, Jenkins-Smith, and Braman 2011). Far from being inclined to 'give in' when our views are criticized, the more usual response is to find ways of rejecting those criticisms. We are, in general, motivated to find ways of rejecting evidence and information that seems to cast doubt on our views (recall Taber and Lodge 2006).

If all this is right, many of us will frequently and predictably fail to conduct our inquiries in the way Roberts and Wood say is required for intellectual autonomy. So intellectual autonomy, at least as they conceive of it, is a goal that many of us will frequently and predictably fail to achieve, at least a fair amount of the time. (How much of the time? Motivated reasoning is well-documented in political cognition, and its influence on our self-conceptions and -understandings is also well-documented. While there is no reason to think its influence extends to all human cognition, it clearly extends quite far.)

This is a persuasive argument against Roberts and Wood's view of intellectual autonomy, especially as it is part of their regulative epistemology that the achievability of goals and ideals matters. Their thought is that a goal cannot regulate our inquiry unless we have a realistic hope of achieving it. But you might respond, even if not on Roberts and Wood's behalf, that the fact that a goal is hard to attain does not mean we should not strive for it. A goal like justice is (very) hard to attain. But this is no reason for thinking that we shouldn't strive to attain it (for a similar point, see Emmet 1994). In my discussion of Carter, I argued that not only is intellectual autonomy hard to attain, but it is also a goal that serves to frustrate our other epistemic goals. We need to sacrifice something for intellectual autonomy, and the sacrifice is often not worth it. I need to show that the same goes for intellectual autonomy as Roberts and Wood understand it.

My argument is based on three claims. First, our tendency to engage in motivated reasoning limits our intellectual autonomy, but we have no reason to think that striving to be intellectually autonomous is an effective way of minimizing our tendency to engage in motivated reasoning. If our tendency to engage in motivated reasoning means we are not intellectually autonomous, trying to be intellectually autonomous will not help because it will not stop us engaging in motivated reasoning.

Now, there is some evidence that we are less inclined to engage in motivated reasoning in certain situations. John Bullock, Alan Gerber, and Seth Hill (2015) report a study where participants were offered small financial incentives for giving correct answers to factual political questions such as 'did the deficit rise under the current administration?'. In their study, there was no correlation between participants' political leanings and their answers, suggesting that they were not engaging in politically motivated reasoning. While there is a debate about what exactly this does (and doesn't) show (Kahan 2016b), it clearly doesn't tell us anything about intellectual autonomy, or show that wanting to be intellectually autonomous makes you less inclined to engage in motivated reasoning.

Second, in Chapter 4 I discussed some strategies science communicators have suggested for 'working around' motivated reasoning. One way this might be done is by framing issues in ways designed not to 'trigger' our ideological biases. Another is by selecting spokespersons who can 'speak to' certain audiences. I argued that, at least if they are used judiciously, these methods need not prevent anyone from developing the capacity for intellectual autonomy. Indeed, they may have the result of making us more intellectually autonomous. What this suggests is that we need some help to develop the capacity for intellectual autonomy. It is not something we can do on our own (more on this in §5.6).

Third, a plausible corollary of this is that, if we refuse this help—if we try to become more intellectually autonomous on our own—we are liable to fail. Not only that, we are also liable to miss out on other epistemic goods. When we engage in motivated reasoning, we tend to end up believing things that support a positive conception of ourselves (when our motive is to protect our self-conception) or align with our political and social values (when our motive is to protect those values). This means that a lot of us will end up forming false beliefs. Arguably, even when we end up forming true beliefs, these beliefs are not justified—we believe the right thing, but we believe it based on the wrong reasons. (More on this in Chapter 8.)

In summary, not only does striving to be intellectually autonomous not negate our tendency to engage in motivated reasoning, but it also exacerbates the bad epistemic consequences of this tendency. In striving to be intellectually autonomous, we run the risk of frustrating our other epistemic goals. Let me reiterate that this does not mean that we should never strive to be intellectually autonomous, whether in Roberts and Wood's or any other sense. It may be that, in certain situations, the sacrifice is worth it. My claim has been that the sacrifice is often not worth it.

In this and the previous two sections, I have argued that even modest views of intellectual autonomy conceive of it in such a way that it is often unattainable. More importantly, 'modest intellectual autonomy' is a goal that often serves to frustrate our other, usually more important, epistemic goals. It may be that there are certain situations in which we should strive to be intellectually autonomous. But it is not the case that, in general, we should strive to be intellectually autonomous. The argument I have developed in these sections is of a piece with the more general argumentative strategy the non-ideal epistemologist uses against the ideal epistemologist, which is to argue that the norms of inquiry and epistemic goals they propose will, far from improving our epistemic situation, often make it worse. The problem with intellectual autonomy is not (just) that it is unattainable. The problem is that trying to be intellectually autonomous often worsens our epistemic situation.

5.5 Intellectual Autonomy and Epistemic Paternalism

In this section, I briefly consider the implications of my argument for epistemic paternalism. In the concluding section, I briefly address the question of how we might develop the capacity for intellectual autonomy.

Let me quickly recap what I said in Chapter 4. When someone's choices or actions are interfered with, this interference counts as paternalistic if it is justified by reference to the good of the person being interfered with, but they do not consent to it. For example, if you hide all the chocolate because you want to stop me from eating it and you do this because you think it is bad for me to eat chocolate, then you are being paternalistic. We can talk of epistemic paternalism where the interference is with how someone conducts their inquiries and the good in question is their epistemic good—how much they know, how many true beliefs they have, and so on. To use our earlier example, if a judge withholds information about the defendant from the jury, and they do this because they think that otherwise the jury will form a false impression about the likelihood that the defendant is guilty, then they behave in an epistemically paternalistic way towards the jury.

As we saw in Chapter 4, one standard objection to paternalism is that it infringes on our autonomy. In the case of epistemic paternalism, the objection is that it infringes on our intellectual autonomy. Intellectual autonomy, as we have seen, can be understood as the ability to control the shape of our inquiries. We can distinguish between two forms of the objection, one of which is stronger than the other. In the strong form, the objection is that (epistemic) paternalistic interference is *always* unjustified because it infringes on our (intellectual) autonomy. In the weaker form, the objection is that there is a standing presumption against paternalistic interference because it infringes on our autonomy. While the stronger form of the objection rules out the possibility of any form of justified paternalistic interference, the weaker form of the objection is consistent with the thought that some forms of paternalistic interference are justified, although there always needs to be a good reason for it.[2]

In this chapter, I have argued that, at least as Carter or Roberts and Wood conceive of it, intellectual autonomy is not something that we always have good reason to strive for. First, it is often unattainable anyway. Second, in striving to be intellectually autonomous, we run the risk of not securing our other epistemic goals. This means that the stronger form of the objection against paternalistic interference is a non-starter.

What about the weaker form? Notice that the weaker form of the objection does not necessarily show that paternalistic interference is rarely justified. It just shows that it is only justified when there is a good reason for it. The argument of this chapter shows that there often is a good reason for epistemically paternalistic interference. I have argued that intellectual autonomy is often in tension with other epistemic goods. If there is no reason to regard intellectual autonomy as more valuable than these other goods, and plenty reason to regard it as less valuable, then it is hard to see why the fact that paternalistic forms of interference can infringe on intellectual autonomy is a weighty reason against such forms of interference. If this is right, then the weaker form of the objection may stand in that there is a presumption against (epistemically) paternalistic interference. But it has little bite because this presumption is often defeated. This chapter therefore amplifies the argument of the previous chapter. Where the previous chapter argued that science marketing need not infringe on intellectual autonomy, this chapter has argued that intellectual autonomy is often not particularly valuable in the first place. Even if science marketing does infringe on our intellectual autonomy, this need not be an objection to it.

The reader, especially the reader who regards intellectual autonomy as valuable, might want to press back here. You might hold that it is up to individuals to decide

---

[2] It is unclear who, if anyone, defends the view that paternalistic interference is always unjustified. This position is often attributed to John Stuart Mill (e.g. Arneson 1980), though there is some debate about whether he really held it (Riley 2018). But the issue is complicated by the fact that, in the literature on paternalism, there is often no agreement about what paternalism is, or about whether examples where it clearly seems justified to interfere with someone's choices or actions for their own good are really examples of paternalism (Scoccia 2018). Because I take the argument of this chapter to show that the strong form of the objection is untenable, at least as it applies to intellectual autonomy, I do not need to get into this here.

whether intellectual autonomy is more valuable for them than the goods I have argued would need to be sacrificed for it. You might also suggest that it is at least not obvious that everyone would prefer to be manipulated, subjected to marketing, or nudged, if the alternative is to hold on to false beliefs about issues like global warming.

It may be that some would prefer holding on to false beliefs, whether about global warming or anything else, to sacrificing their intellectual autonomy. But the crucial question is whether this preference is justified. A preference for intellectual autonomy is only justified if intellectual autonomy really is more important than other epistemic goods. But this is exactly what is at issue. I have been arguing that there is no reason why we should regard intellectual autonomy as more valuable, especially considering that it is often in tension with these other epistemic goods. Unless and until such a reason can be given—and the fact that some might attach special value to intellectual autonomy does not really constitute a good reason— my argument in this chapter stands, as does my defence of certain forms of epistemic paternalism.

5.6 Becoming Intellectually Autonomous

For all I have said, it may well be that intellectual autonomy is prima facie a good thing. I want to finish by considering whether there might be ways of developing it that do not involve sacrificing other epistemic goods. I want to suggest that some forms of paternalistic interference can help us develop the capacity for intellectually autonomous deliberation and thought.

We can start with a remark from Dan Kahan that I discussed in Chapter 4:

It would not be a gross simplification to say that science needs better marketing. Unlike commercial advertising, however, the goal of these techniques is not to induce public acceptance of any particular conclusion, but rather to create an environment for the public's open-minded, unbiased consideration of the best available scientific information. (2010, p. 297)

As I noted, Kahan overstates his point when he says that the goal of science marketing is not to induce acceptance of particular conclusions. It clearly is. But, as I also noted, Kahan is getting at the intriguing idea that there need not be a tension between doing things that are likely to lead to public acceptance of particular conclusions and trying to facilitate engagement with the best available scientific information and evidence. This is because, as Kahan is implicitly assuming, when you engage with the best available evidence in an open-minded and unbiased way, you will conclude that global warming is real, that humans cause it, and that it poses a real danger to humanity.

As we have seen in this chapter, we often do not consider evidence in an open-minded or unbiased way. Rather, we come to it with existing views, and with the motive of preserving those views. So Kahan's claim is that, properly used, science marketing techniques can help create an environment in which individuals properly engage with the evidence for and against scientific theories. It is not much of a step from this to the conclusion that these techniques can help create a more intellectually autonomous public.

To see how this might work in a little more detail, consider one of the strategies I discussed in Chapter 4: prebunking. When you 'prebunk' misinformation, you try to refute it before it can be taken on board. For example, you might present someone with a common climate sceptical argument along with a refutation of it. Take the claim that human CO_2 emissions are tiny in magnitude compared to natural emissions. Out of context, this might seem to support the conclusion that human emissions are, in the grand scheme of things, unimportant. But the idea is that, in presenting this argument, you supply the required context. For example, you might accompany a presentation of the argument with an explanation of how human CO_2 emissions interfere with the natural carbon cycle, putting it out of balance.

This strategy seems to infringe on your intellectual autonomy because it steers your thought in a particular direction. You are, at the very least, invited to reach a particular conclusion (the climate sceptical argument fails). But the idea behind prebunking is to give people the intellectual tools they need to make proper use of the evidence they have. Consider what you would learn if you were taught in the way the prebunking strategy suggests. You would not just learn that a particular climate sceptical argument is spurious. You would learn other important facts (how the carbon cycle works) and some more general lessons. For example, you might learn the lesson that to understand the import of a fact—like the fact that human CO_2 emissions are tiny in magnitude compared to natural emissions—you need to know some other facts. Without knowing these other facts, you can easily draw the wrong conclusion. Once you know more of the relevant facts, the import of the fact you started with becomes clearer.

If you take these general lessons on board, then you will be a step closer to being intellectually autonomous. You will be in a better position to deal with new facts that you may encounter, because you will know that, to understand their import, you usually need to know other facts as well. You will also know that, if you do not know enough relevant facts, you may well be wrong about what you take the implications of the facts you do know to be. You will, in short, be closer to having the sort of 'wisdom about knowledge' that, for Roberts and Wood, is an essential component of intellectual autonomy. This invites the surprising conclusion that prebunking—and perhaps other forms of paternalistic interference—can have the result of bringing us closer to the ideal of intellectual autonomy.

In closing, let me just summarize what I have done in this chapter, and in the previous two chapters. Back in Chapter 3, I compared two approaches to the problem of identifying experts. I argued that one approach (Goldman's) is an example of ideal epistemology while the other (Anderson's) is an example of non-ideal epistemology. Specifically, it is a perfect example of the institutional aspect or face of non-ideal epistemology. In Chapter 4, I argued for an extended version of Anderson's non-ideal institutional epistemology that also covers the problem of consequential false beliefs. In Chapter 5, I addressed something which was implicitly assumed in Chapter 4, which is the value of intellectual autonomy. I argued that we often do better not to try and be intellectually autonomous. Taken as a whole, these chapters completed the two tasks I said needed to be carried out to properly defend Anderson's non-ideal approach to institutional epistemology. They have provided a set of suggestions for constructing a better epistemic environment that is evidence-based. They have also dealt with the worry that non-ideal institutional epistemology proposes a problematic form of infringement on our intellectual autonomy. The result has been a vindication of non-ideal institutional epistemology.

6

The Obligation to Engage

In this chapter and the next, I shift my attention to our obligations and responsibilities as inquirers. This chapter develops an account of one specific obligation (an obligation to engage with challenges to your beliefs). On this account, whether you are under this obligation to engage depends on aspects of your social situation such as your social identity or role. In the next chapter, I argue that epistemic agency and responsibility are themselves socially situated. Taken as a whole, these chapters illustrate the second aspect or face of non-ideal epistemology, which is a picture of epistemic agents or inquirers as more deeply socially situated than is typical in much of contemporary social epistemology.

When we enter into any inquiry, we take on certain obligations and responsibilities.[1] Some of these obligations pertain to the gathering of evidence. We need to gather evidence, and we need to gather enough of the right sort of it before we can reach any conclusions (Flores and Woodard Forthcoming; Hughes forthcoming). Others pertain to the direction and objects of our inquiry. We need to inquire into the things that matter, and we need to do so in a proper and sensible manner. Moreover, we need to be cognizant of the fact that our judgements as to what matters (and what does not matter) may reflect our biases and prejudices (Anderson 1995). Still other obligations pertain to our interactions with other inquirers. Some have argued that we have a duty to object to claims that we take to be false, unwarranted, or harmful (Lackey 2020). Others have argued that we have a more general duty to 'speak our mind' about matters that are of importance to us (Joshi 2021).

---

[1] You might wonder if these are epistemic obligations, moral obligations, or some other species of obligation. This is an interesting question, but it is not one that I address in this chapter, or in this book. It is enough for me that some epistemologists are interested in these sorts of obligations and that they are central to the theory of inquiry. It is, however, important to note that, when I talk about epistemic obligations, I do not mean doxastic obligations (obligations pertaining to our beliefs). Some epistemologists are sceptical about whether we have doxastic obligations (Feldman 1988; McCormick 2020; Wrenn 2007), but I need not take a stand on these issues. It is also worth noting that, because I am not interested in doxastic obligations, the fact that this chapter relies on a broadly consequentialist picture of the norms of inquiry is not as problematic as it might appear. Critics of consequentialism in epistemology typically argue that the consequences of holding a belief are irrelevant to the question of whether we should hold it (e.g. Berker 2013). This may be true, but it is far more plausible to hold that the consequences of following a norm of inquiry (will it lead to gaining more, or more valuable, knowledge?) are part of the reason the norm has a hold on us. This sort of consequentialist position is not committed to the claim that you should believe something because doing so will have good consequences. For a similar point, see Goldman (2015, pp. 139–40).

Non-Ideal Epistemology. Robin McKenna, Oxford University Press. © Robin McKenna 2023.
DOI: 10.1093/oso/9780192888822.003.0006

My focus in this chapter will be on a particular obligation some suppose we have to other inquirers, which is to engage with their challenges to our views and beliefs. The best-known defender of the view that we all have an obligation to engage with challenges to our beliefs is John Stuart Mill. In *On Liberty*, Mill mounts an impassioned and, to many, persuasive defence of freedom of expression. But Mill is not just committed to the view that we should be free to express our opinions. He is committed to the stronger view that we should all actively engage with opinions that run contrary to our own. As Mill put it. 'Truth has no chance but in proportion as every side of it, every opinion which embodies even a fraction of the truth, not only finds advocates, but is so advocated as to be listened to' (2011, p. 94). For Mill, it is not enough that people are free to 'give voice' to their opinions. They must be 'advocated as to be listened to'. We must *engage* with what they have to say. On the Millian picture, this engagement is supposed to, albeit only in the long run, have certain distinctively epistemic benefits. It will lead to a better understanding both of what is true and of why it is true.

In this chapter, I argue against the Millian picture. I argue that the problem with the Millian picture is that it is a prime example of ideal epistemology. In particular, it is a prime example of (an epistemological analogue of) Rawlsian full compliance theory. I also argue that viewing the Millian picture this way helps to explain both its attractions and its deficiencies. It helps to explain its attractions because, as a piece of ideal epistemology, it is compelling: an idealized version of debate may well bring several epistemic benefits. But it also helps to explain its deficiencies because it is only when we recognize that the Millian picture relies on these idealizations that the deficiencies become clear.

Once we eschew these idealizations, a different, non-ideal, picture of our obligations as inquirers emerges. On this picture, the nature and extent of our obligations depends on whether we can expect other inquirers to satisfy their obligations to us. There are some inquirers who can expect other inquirers to not satisfy their obligations towards them and this has implications for what their obligations are. It may be that someone in this position has no obligation to engage with certain challenges to their beliefs even though others, who are differently positioned, are under this obligation.

Here is the plan. I start by outlining Mill's argument that we all have an obligation to engage with challenges to our beliefs (§6.1). As it turns out, Mill gives two arguments in *On Liberty*. The first, which is familiar, is the argument I have already briefly discussed: we all have an obligation to engage with challenges to our beliefs because this is the best way of securing certain epistemic benefits. The second, which is less familiar, is that it is only by engaging with challenges to our beliefs that we gain or retain the right to our beliefs. Because this argument is not well developed by Mill, I look to some recent work by Quassim Cassam on intellectual vices to flesh it out in more detail (§6.2). I then raise some objections to Mill's and Cassam's arguments (§§6.3–6.4). After developing my diagnosis of where the Millian picture goes wrong (§6.5), I finish by comparing my argument against Mill to some recent work by Nathan Ballantyne, Jeremy Fantl, and Amia Srinivasan (§6.6) and by sketching what a non-ideal theory of our obligations as inquirers looks like (§6.7).

6.1 Mill on the Obligation to Engage

In this section, I take a closer look at Mill's argument that we all have an obligation to engage with challenges to our beliefs. We can start with this famous passage from *On Liberty*:

> [T]he peculiar evil of silencing the expression of an opinion is that it is robbing the human race; posterity as well as the existing generation; those who dissent from the opinion, still more than those who hold it. If the opinion is right, they are deprived of the opportunity of exchanging error for truth; if wrong, they lose, what is almost as great a benefit, the clearer perception and livelier impression of truth produced by its collision with error.   (2011, p. 33)

For Mill, the problem (the 'evil') with restricting freedom of expression is that it has bad *epistemic* consequences (MacLeod 2021). To use slightly different language, restricting freedom of expression diminishes the 'stock' of true beliefs that are in circulation, whether because it prevents someone from giving voice to what is true, or because it deprives us of the chance to revisit why we think what we do and develop a better understanding of the reasons supporting our beliefs. On the other hand, Mill thought that having minimal restrictions on freedom of expression would bring certain epistemic *benefits*. It would increase the stock of true beliefs in circulation and facilitate the development of a better understanding of the reasons supporting our beliefs.

Mill's argument for freedom of expression has, to put it mildly, been the subject of heated debate. But no matter your views on whether it succeeds or not, merely *permitting* the free expression of opinion need not secure any of the benefits adverted to by Mill. Merely permitting the expression of an opinion does not mean that anyone is going to listen to it. If nobody is listening, it does not matter how right you are or what might be gained from engaging with you. Mill recognized this, and endorsed the logical conclusion that we need to *engage* with what others say:

> [E]ven if the received opinion be not only true, but the whole truth; unless it is suffered to be, and actually is, vigorously and earnestly contested, it will, by most of those who receive it, be held in the manner of a prejudice, with little comprehension or feeling of its rational grounds. And not only this, but...the

meaning of the doctrine itself will be in danger of being lost, or enfeebled, and deprived of its vital effect on the character and conduct: the dogma becoming a mere formal profession, inefficacious for good, but cumbering the ground, and preventing the growth of any real and heartfelt conviction, from reason or personal experience.   (2011, p. 95)

We gain the benefits of freedom of expression only if the opinions that people are free to express are 'vigorously and earnestly contested'. Mill's view is that we can only secure the benefits of freedom of expression if we all engage with what other people have to say. Engaging will be particularly beneficial when what they have to say *challenges* our own beliefs. We can extract this argument from Mill.

(1) Engaging with challenges to our beliefs is the best way of securing certain epistemic benefits, that is, increasing our stock of true beliefs and developing a better understanding of the reasons supporting our beliefs.
(2) If engaging with challenges to our beliefs is the best way of securing these benefits, we all have an obligation to engage with challenges to our beliefs.
(3) We all have an obligation to engage with challenges to our beliefs.

Let me note two things about this argument. First, we need to distinguish between two versions of the first premise. On the first version, the claim is that engaging with challenges to their beliefs is the only way for an individual to improve their own epistemic situation, whether by increasing their own stock of true beliefs or developing their own understanding. On the second, the claim is that a society where everyone engages with challenges to their beliefs—where there is a 'culture of robust critical discussion'—will have more true beliefs and a better understanding of the reasons supporting those beliefs than a society without such a culture. While these claims are compatible, they are distinct. Engaging with challenges to your beliefs may secure these benefits for society at large without securing them for you, and you may secure these benefits without benefitting society at large. While the passages I have quoted seem to suggest that Mill favoured the second, societal, version of the premise, I consider the prospects for both versions in this chapter.

Second, the argument implicitly assumes a form of consequentialism: an obligation to challenge comes from the beneficial consequences of doing so. This is not surprising, given that Mill was a consequentialist. But it is worth saying a little about consequentialism in the context of inquiry epistemology, which is what we are concerned with here. Consequentialist theories in ethics assume some shared conception of the good and then try to figure out which actions will maximize that good. Similarly, consequentialist theories in inquiry epistemology assume some shared conception of what inquiry aims at (here, truth and a sort of understanding) and then try to figure out what inquirers need to do to attain those aims (see Goldman 1999 on 'veritistic' social epistemology). Because I endorse some form of consequentialism within inquiry epistemology, I am not going to take issue with this in what follows. But it is worth mentioning that one way of rejecting Mill's argument is just by rejecting the consequentialist assumptions on which it is based.

The argument so far is based on considerations about what is likely to secure certain epistemic benefits. But there are places in *On Liberty* where Mill has a slightly different sort of argument in mind. Consider this passage:

> He who knows only his own side of the case, knows little of that. His reasons may be good, and no one may have been able to refute them. But if he is equally unable to refute the reasons on the opposite side; if he does not so much as know what they are, he has no grounds for preferring either opinion. The rational position for him would be suspension of judgment, and unless he contents himself with that, he is either led by authority, or adopts, like the generality of the world, the side to which he feels most inclination.   (2011, p. 67)

Here, Mill is saying something about what is required for a belief to be rational or, as contemporary epistemologists would put it, justified. His claim is that someone who is unable to refute reasons that have or might be offered against their belief is not justified in holding on to their belief. Someone in this position should suspend judgement—they should abandon their belief. It therefore seems reasonable to read Mill as being committed to the view that any inquirer needs to engage with challenges to their beliefs because, if they cannot, those beliefs are unjustified. This second argument seems different from the first, though they are related. One benefit for the individual in engaging with challenges to their beliefs may be that, in doing so, they gain or retain justification for those beliefs.

Mill does not develop the second argument in the same detail as the first so some of the details are unclear. Do you really need to be able to refute all the reasons that might be offered against your belief for it to be justified? This seems an extremely high bar and it is unclear how often we will meet it. But how high a bar it is depends on what 'refuting the reasons on the opposite side' requires. If it requires having a ready rebuttal of any reasons that might be offered, then it is an extremely high bar. On the other hand, if it is enough to have some evidence that speaks against whatever reasons might be offered against your belief, then perhaps the bar can be met with relative ease.

Because Mill does not discuss these issues, it is hard to evaluate his argument. Fortunately, in his 2019 book *Vices of the Mind*, Quassim Cassam develops a similar argument, though as part of a discussion of intellectual vices and not by reference to Mill. It is worth looking at what Cassam says to see if we can extract a plausible version of the second argument from his work.

6.2 Cassam on the Obligation to Engage

We can start with Cassam's account of the intellectual vices. Cassam defends a view he calls 'obstructivism', according to which epistemic or intellectual[2] vices are 'blameworthy or otherwise reprehensible character trait[s], attitude[s], or way[s] of thinking that systematically [obstruct] the gaining, keeping or sharing of knowledge' (2019, p. 23). For Cassam, character traits like closed-mindedness, dogmatism, and gullibility are intellectual vices in part because they systematically produce bad epistemic outcomes: they systematically obstruct the gaining, keeping, or sharing of knowledge. But for a trait to be a vice it must also be something for which the possessor is 'blameworthy or otherwise reprehensible'. This means that a character trait like forgetfulness, which often gets in the way of knowledge but is not (generally) a trait that you are blameworthy or reprehensible for having, doesn't qualify as an intellectual vice on Cassam's account (see Battaly 2019).

At least on my reading, Cassam gives two distinct reasons why intellectual vices systematically 'get in the way' of knowledge. The first, which runs through his entire book, is that the intellectually vicious agent is going to end up not knowing things because of their viciousness. Cassam makes excellent use of 'case studies' where individuals did not recognize important truths due to their various intellectual vices.

The second way, which is the focus of chapter 6 of his book, is that intellectual vices can get in the way of knowledge by undermining our right to our beliefs. Imagine that Catriona takes herself to know that her partner is honest. However, her friends constantly tell her that her partner is a fraudster who is trying to steal all her money. Further, her friends present her with compelling evidence that her partner is a fraudster. But, because Catriona is closed-minded and dogmatic when it comes to her partner's character, she rejects this evidence out of hand.

Whether or not her partner is a fraudster, Catriona is being closed-minded and dogmatic here. Even if Catriona is not a particularly closed-minded or dogmatic person, it is fair to say that, as far as her partner's character is concerned, her thinking displays both these vices. On Cassam's view, someone's thinking on a given occasion can display a vice (a 'thinking vice') even though they lack the relevant underlying character trait (a 'character vice'). Now imagine that, as it turns out, Catriona is right and her friends are wrong—Catriona's partner is honest, and all the evidence against him is misleading. What Cassam would say about Catriona is that, even though her belief is true, she does not have the right to it. She does not have the right to her belief because she has ignored evidence against it, and she ignored evidence against it due to her intellectual vices.

---

[2] I use the term 'intellectual vice' rather than Cassam's preferred 'epistemic vice' in this chapter because in the next I talk about intellectual vices, and it made sense to prefer terminological uniformity over fidelity to Cassam's usage.

Cassam's thought then is that, while intellectual vices can lead us to ignore evidence against our beliefs, they can never *justify* us in ignoring evidence against our beliefs, even if the evidence is in fact misleading. As he puts it:

> [I]f I have encountered what purports to be conclusive evidence against P, and I have no idea how to refute that evidence, then it seems that I no longer have the right to be confident that p [and if] I no longer have the right to be confident that P, then I no longer know P.   (2019, p. 116)

Cassam's view is that you must engage with—and be able to refute—challenges to your beliefs to retain the right to them. (Note that this is a necessary, not a sufficient, condition for having the right to your beliefs. Imagine there is evidence against your belief that you should have been aware of, but nobody makes you aware of—your belief could be challenged but, in fact, it has not been challenged. I see no reason why Cassam could not say that you also do not have the right to this belief. Inquirers have many obligations, including an obligation to be diligent in the gathering of evidence.)

More generally, Cassam emphasizes that there are certain things you must do to qualify as a 'knower' (someone who has knowledge). Knowers have responsibilities, and one of those responsibilities is not to dismiss evidence against and challenges to their beliefs in the absence of good reasons for doing so (2019, p. 119). He uses several examples to illustrate his view.

Imagine you believe that global temperatures are rising due to human activity but are aware of evidence against your belief—perhaps you are aware of some of the claims made by global warming sceptics. For Cassam, you have the responsibility to engage with their claims and try to refute them. If you do not do so, you may remain as confident in your belief as you were before, but you lose your right to be confident. Or, to take another example, imagine you believe that the 11th September attacks on the World Trade Center were carried out by al-Qaeda but are aware of evidence against your belief—perhaps you are aware of some of the claims made by '9/11 truthers'. On Cassam's view, you have the responsibility to engage with their claims and try to refute them. Again, if you do not do so, you may remain as confident in your belief as you were before, but you lose your right to be confident.

For my purposes, Cassam's key claim is that, to 'retain the right' to a belief, you must engage with challenges to it. In the next two sections, I will be arguing against this claim. (I compare my arguments to some related arguments in §6.6.) But the reader might already be wondering whether Cassam is asking too much of inquirers. Can we really be expected to go out and do research into climate change and 9/11 to retain the right to our beliefs about these things? While I am sympathetic to this worry—it is exactly the sort of thing a non-ideal epistemologist would worry about!—I am inclined to think that what Cassam says about it is satisfactory. He says this:

Is this asking too much of [you]? It is certainly at odds with the notion that [you] can retain her right to be confident that P, and [your] knowledge that P, without having to lift a finger ... If asking 'too much' of [you] means requiring [you] to do something to protect [your] knowledge then it's true that too much is being asked of [you]. This interpretation of what it is reasonable to expect from knowers is too undemanding. In many cases the amount of time and effort required to discharge this epistemic obligation is not great, and certainly not beyond the reach of anyone with access to the internet and an attention span of more than five minutes.   (2019, p. 119)

It may be that Cassam is too optimistic about how much effort is needed. If you think that knowing (that p) requires the ability to deal competently with evidence against p, it is hard to see how five minutes' research could be enough to acquire this ability. (If it is enough to acquire it, it is hard to see why you lacked the ability before the five minutes' research.) Still, if you have a view like Cassam's, a middle ground must be struck between placing demands on inquirers that are too excessive (demands that are so hard to meet that they will rarely be met) and demands that are too minimal (demands that are so easy to meet that they are often met already). I am not going to argue that Cassam has struck this balance, but I am not going to argue that he has failed to do so either.

Putting this together, Cassam thinks we all have an obligation to engage with challenges to our beliefs because doing so is required to gain or retain the right to our beliefs. This fits with Mill's suggestion that someone who is unable to refute reasons against their beliefs does not have the right to them. But, unlike Mill, Cassam does not claim that we need to be able to refute any reasons that might be offered against our beliefs. We just need to be able to refute reasons against them of which we are aware. So Cassam offers us a more fleshed out elaboration of Mill's idea that we have an obligation to engage with challenges to our beliefs because it is only by doing so that we can justified in holding on to them.

6.3 The Obligation to Engage in Inhospitable Environments

We have seen that Mill offers two arguments that we all have an obligation to engage with challenges to our beliefs, the second of which is developed in more detail, and in a more plausible direction, by Cassam. In this and the next section, I say why I think both arguments fail. In this section, I consider cases where epistemic agents are in inhospitable epistemic environments. In the next, I consider cases where, while the environment may not be inhospitable for everyone, it is inhospitable for certain epistemic agents. I argue that both sorts of cases cause problems for Mill's and Cassam's arguments.

My argument in this section is based on Heather Battaly's insightful 2018 paper 'Can Closed-Mindedness be an Intellectual Virtue?'. Battaly asks us to imagine an inquirer who is in an epistemically inhospitable environment—an environment where misinformation abounds, and where many inquirers have false beliefs. Let us call our inquirer Laurie. Despite her environment, Laurie has managed to form many true beliefs about what is going on around her. Further, she has arrived at these true beliefs by considering evidence that she has uncovered with a lot of effort and at great personal cost. It seems likely that Laurie is going to have certain intellectual virtues. Getting things right in a world where everyone else has it wrong requires real independence of thought and mind. But you might think Laurie is also going to have certain intellectual vices. As Battaly puts it:

> When a knowledge-possessing agent is stuck in an epistemically hostile environment, surrounded by falsehoods, incompetent sources, and diversions, closed-mindedness about options that conflict with what she knows will minimize the production of bad epistemic effects for *her*.   (2018, p. 39)

The thought is that, if Laurie starts to consider the possibility that she is the one who is wrong, then she might end up losing the knowledge she has fought so hard to gain. It seems like a refusal to engage with challenges to her beliefs—perhaps even a refusal that amounts to closed-mindedness or dogmatism—is going to be required for Laurie to retain her knowledge.

To see why, we can add a little to the case. Laurie has regular interactions with other people. Many of these interactions vividly remind her that she is very much in the minority. Many intelligent people disagree with her about certain things and moreover have arguments and evidence to support their views. If Laurie remains open to the possibility that she is wrong and they are right, or if she tries to engage with these arguments and evidence, she runs the very real risk of convincing herself that she really is wrong. But, if she 'closes herself off', so to speak, she may be able to avoid worsening her epistemic position in these ways. Battaly's view is that, if an intellectual vice is something that gets in the way of knowledge (as it is for Cassam), then Laurie's closed-mindedness (if that is what it is) cannot be intellectually vicious because, far from getting in the way of knowledge, it enables her to retain it.

Now, Battaly's aim in her paper is to argue that, in situations like this, closed-mindedness can be an intellectual virtue. It may well be that what Battaly really shows is that behaviour that would be closed-minded in a normal situation is not really closed-minded in Laurie's abnormal situation. But my interest is not so much in questions of intellectual vice but in whether Laurie (or someone in a situation like Laurie) has an obligation to engage with challenges to her beliefs. It is plausible that she does not. But, if this is right, then Mill's (and Cassam's) arguments must fail, at least in environments like Laurie's.

Let us start with Mill's first argument. Mill's thought was that engaging with challenges to our beliefs is the only way of securing certain epistemic benefits. What cases like Laurie's show is that, at least in inhospitable epistemic environments, engaging with challenges to our beliefs will not secure these epistemic benefits. Engaging with challenges certainly will not secure these benefits for Laurie, for the reasons I have laid out. But it is hard to see how Laurie's engaging with challenges will secure these benefits for her society either.

Perhaps the best that can be said is that, if there were real freedom of expression in Laurie's society (I have built it into the case that there currently isn't) *and* everyone started to engage with the views expressed by others, *then* engaging with challenges to her beliefs might bring some benefits, both for Laurie and for everyone else. But, even if this were true, it would not be grounds for saying that Laurie has an obligation to engage with challenges to her beliefs in her society as it is right now. It would, at best, be grounds for saying that, if Laurie's society were better—if it were less epistemically inhospitable—she would have this obligation.

Turning to Mill's second argument, the idea was that we all have an obligation to engage with challenges to our beliefs because otherwise we lose the right to them. We can ask if, when Laurie refuses to engage with those who challenge her beliefs, she loses her right to the beliefs she is trying to protect. It is, at least on the face of it, plausible that, if Laurie manages to 'shut herself off', she retains her right to these beliefs. But can I say anything to make it more than just initially plausible? I want to offer two ways of thinking about Laurie that should bolster any intuition you might start with.

First, we can view Laurie as having 'latched on' to some important truths. What these truths are depends on how we fill in the case. Perhaps Laurie lives in a nightmarish dystopia like the Republic of Gilead from Margaret Atwood's *The Handmaid's Tale*. Or we can drop the fiction and place her in Nazi Germany, Stalinist Russia, or Jim Crow-era America. Either way, the idea is that we can view Laurie as having latched on to some important truths about what her world is like—how it really works, what is wrong with how it works, and so on. She has latched on to these important truths even though she is unable quite to explain how she has done so, or to deal with the impressive-sounding arguments that might be offered by those who disagree with her. Her beliefs are responsive to the right considerations, even though she is not able to convince anyone else that this is so. In my view, the mere fact her beliefs are responsive to the right considerations is a reason to think that she retains the right to them, even if she can't deal with challenges to those beliefs (cf. Srinivasan 2020).

You might reply that the considerations to which our beliefs ought to be responsive include challenges and responses to our beliefs. But this would be to insist that you need to deal with challenges to your beliefs to have the right to them, which is exactly what is at issue. This suggests that there might be a deeper disagreement here over the requirements on justified belief and knowledge. But if Mill's (and Cassam's) argument that we have an obligation to engage with challenges to our beliefs relies on a substantive view of justification and know-ledge, then that view needs to be spelled out and discussed in detail. Moreover, this view needs to deal with the fact that it is not obvious that, if Laurie manages to 'shut herself off' and not deal with any challenges to her beliefs, those beliefs are thereby unjustified. (Those who are well versed in contemporary epistemology will recognize that I am effectively taking a stand here in the debate between epistemic externalists, who typically think that being reliably responsive to the right considerations is sufficient for justified belief, and epistemic internalists, who typically do not. More on this in §6.6.)

Second, in thinking about Laurie's situation we can build on some work by Miranda Fricker on what we should say about agents who do not pick up on shifts and deficiencies in routine moral thinking. In her 2010 paper 'The Relativism of Blame and Williams' Relativism of Distance', Fricker argues that the appropriate emotional response to someone who continues to defend the status quo, even though deficiencies in it are becoming apparent and some are starting to question it, is a form of *disappointment*. Disappointment is like blame in that it involves the imputation of responsibility (they are responsible for not shifting too) but weaker in that it accepts that this responsibility is diminished (those who are shifting are the exception not the norm).

Building on Fricker, we can ask what the proper emotional response is to someone who is quick to pick up on deficiencies in routine thinking. I am not sure what the best term for this response is, but it is likely a species of *admiration*. Whatever we call this response, Laurie is a fitting target of it. She has managed to respond to the right considerations in forming her beliefs, even though there was so much misleading evidence that might have led her to ignore these consider-ations. This reflects well on her ability to identify the right considerations, and to be swayed by them. But, if we can view Laurie as the proper target of this form of admiration, it is hard to see how we can also view her as having failed to satisfy her obligations and responsibilities as an inquirer. She could have done better still (imagine she could explain why everyone else is wrong). But it is one thing to say she could have done better and quite another to say she has failed to discharge her obligations.

Of course, this is not a 'knock down' argument that Laurie was under no obligation to engage with these challenges. It is possible that Laurie has performed very admirably in some respects (she has figured things out) yet she has not fulfilled some obligations she was under. But pointing out how well she has performed, intellectually speaking, puts the onus on someone who wants to insist that, still, Laurie has violated some obligation or other.

In closing let me comment on a limitation of my argument in this section. I have argued that agents in inhospitable epistemic environments do not have an obligation to engage with challenges to their beliefs. But you might object that we do not inhabit (such) an inhospitable epistemic environment. What about agents in our world? In the next section, I consider what to say about the obligations of agents for whom the environment is inhospitable, even if the environment is not inhospitable in general.

6.4 The Obligation to Engage and Epistemic Exclusion

Consider these two cases:

Nadja is a postgrad on a large MA programme. In her seminars, she notices that the male students (of whom there are many) speak over the female students (of whom there are few), take credit for what female students have said, and dominate discussion. Nadja forms various beliefs about the gender dynamics of her seminars—male students dominate, they take unfair credit, and so on. She speaks to other students in her seminar about the gender dynamics. While the other female students agree with her, all the male students reject her concerns out of hand. More than that, they manage to convince her that her concerns were baseless. But her concerns were not baseless—male students do indeed dominate classroom discussions.

Sarah works in a large organization. She has spoken to several women in the organization who have told her that a senior male member of staff regularly behaves inappropriately (e.g. he makes sexist remarks). She forms various beliefs about this senior male staff member—that he is sexist, that he is unprofessional, and so on. She speaks to some of her male colleagues. All the male colleagues take the view that either the women Sarah spoke to misunderstood what had happened, or the allegations were deliberate fabrications. More than that, they manage to convince her that their view of the matter is correct. But Sarah's original view of the matter was correct. This male staff member does indeed regularly behave inappropriately.

In each case, a subject initially forms some true beliefs then abandons those beliefs in the face of challenges from their colleagues. It is plausible that both Nadja's and Sarah's initial beliefs (their beliefs before they were challenged) qualify as knowledge. Nadja's beliefs about the gender dynamics in her classroom are based on her observation of the gender dynamics. If Nadja is reasonably perceptive, then there is no reason to deny that she can gain knowledge through her observation. Sarah's beliefs about this male staff member are based on testimony. If Sarah is discerning in accepting testimony, then there is also no reason to deny that she can gain knowledge through this testimony. Absent reasons to be sceptical about the possibility of knowledge through observation of social dynamics or testimony, then, both Nadja and Sarah's original beliefs qualified as knowledge. But, when Nadja and Sarah engage with challenges to their beliefs, they end up abandoning their beliefs. So, while their original beliefs qualify as knowledge, by considering challenges to their beliefs they end up losing knowledge.[3]

My claim is that neither Nadja nor Sarah was ever under an obligation to engage with these challenges. These cases therefore cause further problems for Mill's and Cassam's arguments. I hope this claim is plausible enough on its face. Still, even if the reader does think it is plausible, I need to explain why it is plausible. To this end, I want to introduce a framework for thinking about these cases. This framework is taken from Kristie Dotson's influential work on epistemic exclusion, epistemic oppression, and silencing (see Dotson 2011, 2014, 2018).

According to Dotson, epistemic exclusion occurs when someone is prevented from contributing to the production of knowledge. The knowledge in question might concern a matter of public import (e.g. the use of racial profiling by the police). Or it might concern some matter that is primarily of import to the individual in question (e.g. an experience they recently underwent). Dotson distinguishes between being prevented from contributing to the production of knowledge in a particular situation, which is unfortunate but not indicative of a pattern of exclusion, and epistemic oppression, which 'refers to persistent epistemic exclusion that hinders one's contribution to knowledge production' (2014, p. 115). When someone is persistently excluded from the means of knowledge production, we can say that they experience epistemic oppression (2014, p. 116). We can expect that those who experience epistemic oppression will typically be members of social groups that are often excluded from social practices due to their social identity (their age, class, ability status, gender, race, sexual orientation, etc.).

---

[3] Let me briefly address two worries about this claim. I address these worries in a footnote so as not to disrupt the flow of the argument. First, you might wonder whether I am relying on any substantive claims about defeat and the situations where gaining a defeater for a previously known belief can lead you to lose knowledge. But all I am relying on is the uncontroversial claim that, if you do not believe that p, then you do not know that p. (I may be able to make use of an even more uncontroversial claim: even if there are some cases in which you can know that p without believing that p, this is not one of them.) Because Nadja and Sarah abandon their beliefs after considering challenges to them, they no longer have knowledge after considering these challenges. Second, you might deny that either Nadja or Sarah ever had knowledge in the first place because their original beliefs were too fragile. The thought would be that, if Nadja and Sarah can be so easily persuaded that their original beliefs were mistaken, they did not know in the first place. But, if they did not know in the first place, engaging with challenges to their beliefs did not lead to them losing any knowledge. It is important to highlight that, at least as I am construing the cases, neither Nadja nor Sarah is overly deferential to the opinions of others. They are not disposed to abandon their beliefs whenever they learn that others disagree with them, for example. They are disposed to abandon their beliefs when their attempts to defend them are met with incredulity and incomprehension. I submit that most of us would be inclined to abandon most of our beliefs if they were met with incredulity and incomprehension on the part of our interlocutors. There is scope to use this in an argument for a form of scepticism. But it would be a radical sort of scepticism— the sort on which many beliefs that we thought were justified turn out not to be. I say a little more about both these worries in §6.6, especially the second, which has something in common with the sort of sceptical arguments presented by Ballantyne (2019).

There are many ways in which you might contribute to knowledge production. You might find out some new information that is of interest to others. You might hit upon a better way of synthesizing an existing body of information. Or you might engage in 'knowledge dissemination'—the transfer of knowledge to others. When we attempt to transfer knowledge to an audience, we are reliant on their good will. As Dotson puts it, 'to communicate we all need an audience willing and capable of hearing us' (2011, p. 238). Silencing occurs when this good will is lacking.

Dotson (2011) identifies what she calls 'practices of silencing' as being of particular importance here. A speaker is subject to a practice of silencing when their audiences are persistently unwilling to or incapable of hearing them due to pernicious ignorance. By 'pernicious ignorance' Dotson means a persistent and harmful form of ignorance about the speaker and the social group to which she belongs. Consider, for example, an unfounded prejudice that people from a certain social group are untrustworthy or prone to exaggeration. If this prejudice is widely shared, then members of this social group will regularly be subjected to practices of silencing—their audiences will be unwilling to or incapable of hearing them due to the prejudice. Because being subject to a practice of silencing means being persistently excluded from the means of knowledge production, those who are subject to these practices experience epistemic oppression.

Finally, Dotson distinguishes between two kinds of silencing: testimonial quieting and testimonial smothering. Testimonial quieting occurs when a speaker is not recognized as a competent epistemic agent (or a 'knower'). A speaker is subject to a practice of testimonial quieting when they are regularly not recognized as a knower due to pernicious ignorance on the part of their audience. For example, a Black woman might regularly not be recognized as a knower by her sociologist peers because of pernicious ignorance on the part of other sociologists (see Dotson's discussion of Collins 2000).

Testimonial smothering, on the other hand, occurs when a speaker limits their testimony to what they think their audience will be able and willing to understand and accept. A speaker is subject to a practice of testimonial smothering when they regularly limit their testimony in this way. For example, a Black person may find their attempts to discuss issues of race with White people typically go badly and develop strategies for avoiding the most contentious topics, or just stop talking about race entirely, as in Eddo-Lodge (2018). The difference between testimonial quieting and testimonial smothering is that, where quieting occurs in the *reception* of a piece of testimony, smothering occurs prior to the testimony being given—it is a form of self-censorship. That said, it is important not to overstate this difference; one of the main reasons why someone might limit their testimony is how their testimony has been received in the past.

With this framework in hand, let us return to Nadja and Sarah and use the framework to highlight the salient aspects of the cases. Both Nadja and Sarah are excluded from the means of knowledge production. Specifically, they are prevented from disseminating important knowledge about classroom gender dynamics (Nadja) and workplace sexual harassment (Sarah) to their colleagues and to the wider world.

Further, this exclusion is not a one-off incident. We can add to the cases that Nadja and Sarah have regularly found that, when they voice concerns of this sort (concerns about the behaviour of male students or colleagues), these concerns are dismissed, and their interpretation of events is questioned. They are dismissed in part because of widespread, pernicious prejudices (e.g. that women are 'overly sensitive about such things') and other forms of ignorance (e.g. about what sexual harassment is). This is a persistent pattern of exclusion of the sort that, for Dotson, qualifies as epistemic oppression. Because both Nadja and Sarah are persistently excluded from the means of knowledge production, they experience epistemic oppression.

This persistent pattern of exclusion is reflected (and in part based on) the reaction of their audiences to their testimony. Their audiences are not willing or able to listen to them properly—to take their concerns seriously—because they are unwilling or just unable to try to understand. This is not a one-off incident but is part of what Dotson would call a practice of silencing. In the past, Nadja and Sarah have not been taken seriously when they have tried to voice concerns about these sorts of matters.

Each case has aspects of both testimonial quieting and testimonial smothering. When Nadja and Sarah initially present their testimony (their interpretations of what is going on in the classroom or the workplace), they are not taken seriously as knowers. They both (plausibly) have knowledge, but they are not recognized as having knowledge; they are subject to a practice of testimonial quieting. Subsequently, they modify their testimony in light of the hostile reaction of their audiences in a way that is typical of testimonial smothering. However, Nadja and Sarah do not just modify their testimony. Ultimately, they end up abandoning their beliefs. There is something more than just testimonial smothering here; testimonial smothering occurs when someone censors their testimony for public consumption, not when they change their own mind about the matter in question.

This illustrates the point, made by Fricker (2007) among others, that not being taken seriously as a knower can lead to serious epistemic harms. One of these harms is that you can end up losing confidence in your ability to obtain knowledge, at least within the domains where your competence has not been recognized. This can happen because, as Dotson highlights, someone who is not taken seriously due to pernicious ignorance is typically persistently excluded from the means of knowledge production and dissemination. The intuitive thought is that, if your testimony on certain topics is regularly rejected, and others often tell you that you are wrong about these things, then you might start to take the possibility that you really are wrong about these topics seriously. Crucially, this can happen whether you are generally wrong about these topics or not, and even if you are not particularly intellectually humble. Indeed, you might think it takes something more like intellectual arrogance to think you are right when everyone around you says you are wrong.

What does this mean for Mill's arguments? We can start with Mill's first argument, which was that engaging with challenges to your beliefs secures epistemic benefits. Engaging with challenges to their beliefs secures no epistemic benefits for Nadja or Sarah. Initially, they had knowledge that they wanted to disseminate to others, but they failed to disseminate it because they were not taken seriously as knowers. Ultimately, they both ended up abandoning beliefs that were held on grounds strong enough to qualify as knowledge. So, not only have Nadja and Sarah failed to provide an epistemic benefit for their wider community, but they also ended up incurring an epistemic cost to themselves because they have lost knowledge they previously had.

It is important to highlight that this is just a particular instance of Nadja and Sarah's persistent exclusion from the means of knowledge production. It is not just that, on this occasion, no epistemic benefits have resulted from their engaging with challenges to their beliefs. It is typically the case that no epistemic benefits result from them engaging, at least with respect to matters about which they are not taken seriously as knowers due to pernicious forms of prejudice and ignorance.

More generally, we need to recognize that the question of your obligations vis-à-vis challenges to your beliefs cannot be separated from the question of whether you are the sort of person who tends to be included in the means of knowledge production or the sort of person who tends to be excluded. On the Millian picture, there are epistemic benefits to be gained from engaging with challenges to your beliefs because doing so fosters a culture of robust criticism and debate—a culture from which, for the most part and in the long run, truth will emerge. But, once we recognize that we do not all participate in these debates on equal terms, this picture looks less attractive. It may be that those who are included in the means of knowledge production will typically benefit both themselves and everyone else from engaging with challenges to their beliefs. But it is harder to see why those who are excluded will benefit themselves, or indeed anyone else, from engaging.

Turning now to Mill's second argument, and Cassam's development of it, recall the idea was that we all have an obligation to engage with challenges to our beliefs because, unless we do so, we lose the right to our beliefs. Is there scope for arguing that, despite all I have said so far, it was only by engaging with challenges to their beliefs that Nadja and Sarah could have retained the right to their beliefs?

We could say that, even though it turns out badly for them, Nadja and Sarah still had the obligation to engage. But this seems implausible for much the same reasons that it seemed implausible that Laurie had an obligation to engage with challenges to her beliefs. Like Laurie, both Nadja and Sarah have managed to latch on to some important truths and they deserve our admiration for having done so. They lack the resources to fully defend their beliefs, at least to their own satisfaction (you might think that they do have the resources to fully defend their beliefs, but the problem is that they do not realize it). They might have done (even) better as inquirers. All else being equal, it is better to be able to deal with challenges to your beliefs than to not be able to do so. But it is hard to see why this should lead us to think that they have failed to fulfil their obligations and responsibilities as inquirers.

This completes my case against the two arguments that we all have an obligation to engage with challenges to our beliefs that I started with. In the next section, I develop a diagnosis of where these arguments—and the Millian picture—goes wrong.

6.5 Full vs. Partial Compliance Theory

In Chapter 2, I outlined Rawls' distinction between full and partial compliance theory. Applied to epistemology, the distinction is between two different ways of approaching the question of what our obligations are as inquirers. On the full compliance approach, we consider these obligations on the assumptions that (i) other inquirers will comply with their obligations, and (ii) our environment is epistemically hospitable. On the partial compliance approach, we consider these obligations without making either assumption. As I argued in Chapter 2, this way of distinguishing between ideal and non-ideal epistemology is best viewed as a special case of a more general distinction (taken from Mills) between an approach to epistemology that involves certain idealizations about inquirers and the environments in which they are embedded and an approach that avoids such idealizations. Still, in the rest of this section, I will work with the epistemological analogues of full and partial compliance theory as they serve my purposes well.

My contention is that the Millian picture, on which engaging with challenges to our beliefs typically brings certain epistemic benefits, goes wrong because it is an example of full compliance theory. First, it assumes that other inquirers will comply with their obligations. At least, a charitable explanation why someone might think that engaging with challenges to our beliefs will bring epistemic benefits (whether for us or everyone else) is that they are ignoring the fact that some of us are persistently excluded from the means of knowledge production. As I argued in §6.4, if you are persistently excluded from the means of knowledge production, the benefits of engaging with challenges that Mill adverts to will likely not materialize. They certainly will not materialize for you (you are excluded). But it does not look like they will materialize for anyone else either.

To develop this point in a little more detail, we can identify an obligation that is violated in cases like Nadja's or Sarah's. As I discussed in the previous section, the basic problem is that neither Nadja nor Sarah is taken seriously as a knower—as an individual with important knowledge to impart to others—due to pernicious forms of ignorance on the part of their audience (e.g. ignorance about what sexual harassment is). Part of not being taken seriously is not being regarded as being as credible as one in fact is (Fricker 2007). Following Fricker, we can say that a recipient of testimony should assign as much credibility to a speaker as they are due. If they are credible, the recipient should regard them as credible; if they are not, the recipient should not regard them as credible. Call this the 'Proportional Weight Requirement': you should proportion the weight you assign to testimony to the actual degree of credibility of the testifier.

Of course, many violations of the Proportional Weight Requirement will be entirely blameless, as when you have good reasons to think that someone is not credible when they in fact are credible, or vice versa. But a blameless violation of a norm or obligation is still a violation; it is just a violation for which you cannot be blamed. More importantly, in cases like that of Nadja or Sarah, the audience's violation of the Proportional Weight Requirement is not obviously blameless. The reason their audience do not take them seriously as knowers is that they are prejudiced, and we typically think you can be blamed for your prejudices (more on this in Chapter 7). At any rate, while violating the Proportional Weight Requirement is not a sufficient condition for perpetuating persistent patterns of epistemic exclusion, it does seem to be a necessary condition for doing so.

Second, the Millian picture also assumes that the environment is epistemically hospitable. At least, a charitable explanation why someone might think that engaging with challenges to our beliefs will bring epistemic benefits is that they are assuming that the environment is (sufficiently) hospitable. As I argued in §6.3, when we look at an agent in an inhospitable epistemic environment it becomes clear that their engaging with challenges to their beliefs will not bring the adverted epistemic benefits.

But—the reader might ask—why is it a problem that the Millian picture is based on these assumptions? One reason it is a problem is just that, as I have argued, the picture is wrong. It tells us that we have obligations that we do not have. But I also think that viewing the Millian picture as based on these assumptions supplies the basis of a diagnosis of where and why the Millian picture goes wrong. To see this, let us consider two worlds:

The Good World: In the Good World, recipients of testimony are exceptionally good at assigning the proper degree of credibility to testifiers. They sometimes make mistakes. But these mistakes are always innocent mistakes. Further, these mistakes are innocent in part because there are no pervasive and pernicious prejudices to the effect that some testifiers are less credible than others due to features of their social identity (their race, gender, etc.).

The Bad World: In the Bad World, recipients of testimony are unbelievably bad at assigning the proper degree of credibility to testifiers. Not only do they often make mistakes, these mistakes are also so common because of pervasive and pernicious prejudices to the effect that some testifiers are less credible than others due to features of their social identity. As such, these mistakes are not typically innocent mistakes. Those who make them are culpable for them.

In the Good World, inquirers (generally) comply with their obligations. In particular, they (generally) comply with the Proportional Weight Requirement—their assessments of the credibility of testifiers are, for the most part, accurate. In contrast, in the Bad World, inquirers (generally) do not comply with their obligations. In particular they (generally) do not comply with the Proportional Weight Requirement—their assessments of credibility are often inaccurate, and moreover they are inaccurate because of prejudice and other forms of ignorance.

Let us consider how an obligation to respond to challenges would play out in both these worlds. It is plausible that, in the Good World, if testifiers energetically engaged with challenges to their beliefs, this would bring epistemic benefits to them and to everyone else. These benefits might include gaining a valuable 'external perspective' on their beliefs, starting fruitful discussions, and an eventual convergence 'on the truth' of the sort envisioned by Mill. Of course, on occasion things will go wrong. The claim is just that, in the long run and for the most part, these benefits will be secured. Consequently, it is plausible—at least if you are attracted to the Millian picture in the first place—that the inhabitants of the Good World are under an obligation to engage with challenges to their beliefs.

In contrast, in the Bad World, it is far from clear that any of these benefits would result from a culture of robust criticism and debate. What valuable external perspective would the inhabitants of the Bad World gain on their beliefs by engaging with challenges to them? Why think that discussions would be fruitful or eventually converge on the truth? Of course, the claim is not that engaging with challenges would never be beneficial in the Bad World. The claim is just that, in the long run and for the most part, it will not be beneficial. Consequently, even if you are attracted to the Millian picture, it is far less plausible that the inhabitants of the Bad World have any obligation to engage with challenges to their beliefs.

As should already be clear, I do not think that our world is particularly close to the Good World. In our world, epistemic oppression, and the persistent patterns of epistemic exclusion that constitute it, are common. Moreover, it is often the same epistemic agents who suffer as a result. But my point here is not just that the Millian picture is wrong because it assumes that our world is more like the Good World than it really is. My point is that we can explain *the appeal* of the Millian picture by putting things in these terms. Whether someone thinks we are under an obligation to engage with challenges to our beliefs will depend on how close they think our world is to the Good World—the closer they think it is, the more likely they are to think that we are under the obligation. It is therefore not hard to see why the Millian picture would be appealing to the kind of ideal epistemologist who ignores the prevalence of distinctively epistemic forms of oppression and exclusion when theorizing about our obligations and responsibilities as inquirers. The problem, of course, is that, if we are interested in what our obligations really are, we cannot ignore the prevalence of epistemic oppression and exclusion.

I have argued that we do not all have an obligation to engage with challenges to our beliefs. I have also suggested that whether we do have an obligation depends on aspects of our social position such as our social identity, which typically track persistent patterns of epistemic exclusion. In the concluding section of this chapter, I sketch some more of the details of this non-ideal picture of our epistemic obligations. But before doing that I want to explain how my argument so far compares to similar arguments that have been offered in the literature. The reader who is not familiar with this work, or who is not interested in how my arguments compare to it, should skip the next section, and jump straight to §6.7.

6.6 Ballantyne, Fantl, and Srinivasan

The three authors I want to compare my arguments with are Amia Srinivasan, Jeremy Fantl, and Nathan Ballantyne. Let me start with Srinivasan and Fantl, both of whom defend arguments and claims that are closely related to my own.

The reader who is familiar with Srinivasan's 2020 paper 'Radical Externalism' might have noticed a similarity between my cases (Nadja and Sarah) and the cases she uses to argue for an externalist view of justification. (This is the view that a belief can be justified due to the simple fact that it is the product of a reliable capacity to form true beliefs about the matter in question, irrespective of whether the believer is aware that her belief is the product of a reliable capacity.) Here is one of her cases in full:

> CLASSIST COLLEGE: Charles is a young man from a working-class background who has just become the newest fellow of an Oxford college. He is initially heartened by the Master's explicit commitment to equality and diversity. The Master assures him that, though the college is still dominated by fellows from elite socioeconomic backgrounds, Charles will be welcomed and made to feel included. Indeed, the Master tells Charles, he too is from a working-class family, and has experienced plenty of discrimination in his time. Charles is confident not only that the college will be a good community for him, but also that the Master is a person of excellent judgment on these matters. Soon, however, a few

incidents disrupt Charles's rosy view of things. At high table, when Charles explains that he went to a state school, a fellow responds with "but you're so well-spoken!" At a visit to the pub, a number of young fellows sing the Eton boating song while Charles sits uncomfortably silent. Finally, Charles hears that the other fellows have taken to calling him "Chavvy Charles." Charles, who has a dependable sensitivity to classism, goes to the Master to report that he has experienced a number of classist incidents in college. Shocked, the Master asks him to explain what happened. But when Charles describes the incidents, the Master is visibly relieved. He assures Charles that none of these are genuinely classist incidents, but playful, innocuous interactions that are characteristic of the college's communal culture. He tells Charles that he is sure that Charles himself will come to see things this way once he gets to know the college and its ways better. And finally, he gently suggests that Charles is being overly sensitive—something to which (the Master goes on) Charles is understandably prone, given his working-class background. Charles is unmoved. He continues to believe that he has faced classist discrimination in the college, dismissing the Master's testimony to the contrary. Charles meanwhile is unfamiliar with the idea of false consciousness—and, in particular, the phenomenon of working-class people who have internalized bourgeois ideology.   (2020, p. 397)

Srinivasan's claim is that, because Charles has a reliable capacity to form true beliefs within a domain (about class dynamics), his beliefs about the class dynamics in his college are justified. Similarly, my claim is that the subjects in my examples (Nadja and Sarah) had justified beliefs (indeed, knowledge) about gender dynamics in the classroom (Nadja) and sexual harassment in the workplace (Sarah) in virtue of a reliable capacity to form true beliefs about these domains.

The difference is that, where Charles is not moved by (misleading) evidence that his beliefs about class dynamics are false, Nadja and Sarah are moved by misleading evidence and, as a result, abandon their initial beliefs. In these examples, I have claimed, Nadja and Sarah were not under any obligation to engage with this misleading evidence, or with the challenges via which it was presented. As far as I can see, there is nothing in Srinivasan's paper to indicate that she also accepts this claim. So, while I rely on some claims that Srinivasan makes, I have made some additional claims. More generally, our aims differ. Srinivasan's aim is to defend an externalist account of justification. I have relied on an externalist view of justification to defend a view of our obligations as they pertain to challenges to our beliefs.

Moving on to Fantl, in his 2018 book *The Limitations of the Open Mind* he defends a view that is similar to my own:

we have to ask whether an obligation to engage open-mindedly derives from the goods—epistemic or otherwise—that open-minded engagement gets for yourself or the larger society. The response to this must involve a demonstration that open-minded engagement, in many cases, does not produce epistemic benefits for yourself or society at large. This, again, is the case I presented in this chapter. It does not produce epistemic benefits for yourself because, in knowing that the counterargument is misleading, you only stand to distort your confidence by being willing to reduce your confidence in response to a misleading counter-argument.   (2018, p. 152)

I have claimed that some of us are under no obligation to engage with challenges to (some of) our beliefs. Similarly, Fantl argues that we sometimes should not engage with arguments against our beliefs. But how does Fantl reach this conclusion? Here is his basic argument:

(1) There are standard situations in which you know controversial proposi-tions and, thus, know that a relevant counterargument is misleading.
(2) If you know, in a standard situation, that a counterargument is misleading, you should not engage with the counterargument.
(3) There are standard situations in which you should not engage with a relevant counterargument against a controversial proposition.

Before I comment on the differences between Fantl's argument and my own, some comments on the argument are in order. First, a 'controversial proposition' is just a proposition about which there is controversy. For example, the propos-ition that human activity is the main cause of global warming is a controversial proposition. A 'relevant counterargument' is just an argument against a propos-ition that you believe. So Fantl is imagining a situation where you know some proposition about which there is controversy (e.g. that human activity is the main cause of global warming) and are aware of an argument against this proposition (e.g. an argument that human activity is not the main cause). His claim is that, in some situations of this sort, you should not engage with this counterargument.

Second, Fantl's argument for the first premise is based on two further claims. The first is that we can have strong enough evidential support for controversial propositions to count as knowing them—that a proposition is controversial does not mean we cannot have excellent evidence that it is true.

The second is that this knowledge can survive learning that the proposition is controversial. Fantl's argument for this second claim is involved, but the basic thought is this. Even if learning there is an argument against one of your beliefs—one which you cannot refute—provides *some* evidence that your belief is false, this evidence may be weak. If you are not a climate scientist, it is no surprise that you cannot find the flaw in climate sceptical arguments that appeal to scientific data and methodologies. Because it is not surprising, the fact that you cannot find the flaw is not much reason to think there is not a flaw.

Third, Fantl's argument for the second premise is also based on two further claims: you should act in accordance with what you know, and the fact that you know an argument is misleading is usually a decisive reason not to engage with it. The first claim is often called the 'knowledge-action' principle, and it has been defended by Fantl in joint work with Matthew McGrath (Fantl and McGrath 2009). The second claim is plausible enough on its face: in standard situations, the fact that you know a counterargument is misleading does indeed seem to be a good reason not to engage.

What are the differences between Fantl's argument and mine? I want to highlight three. First, as I have already highlighted, my argument relies on an externalist picture of justification: the mere fact that a belief is the product of a reliable capacity to form true beliefs within the relevant domain can be sufficient for that belief to be justified (indeed, for it to be knowledge, if it is also true). Fantl's argument does not rely on this picture. Indeed, he claims that his argument is neutral on the externalism/internalism debate (see Fantl 2018, p. 131, ff. 7). This means that, at least as far as readers of a more internalist persuasion are concerned, I would do well to fall back on Fantl's argument. (On the other hand, Fantl's argument relies on substantive principles connecting knowledge and action while mine does not. Make your choice.)

Second, Fantl's view is that, when you know a relevant counterargument (an argument against some controversial proposition you believe) is misleading, you should not engage with it. He seems to think that, if you do not know it is misleading, even though it in fact is misleading, you should engage with it. While my argument in §6.3 only requires the view that Fantl defends (when you know an argument is misleading, you should not engage with it), my argument in §6.4 requires the stronger claim that there are cases where you should not engage with counterarguments even though you do not know they are misleading.

That said, there are different ways in which you might defend this stronger claim. The way in which I have defended it makes it less controversial than it might initially appear. I have argued that, in a case where you do not know that a counterargument is misleading simply because you do not *believe* it is misleading, it may be that you still should not engage with it. That is, you still should not engage in situations where you are *in a position to know* that the counterargument is misleading but do not in fact know because you are not sufficiently confident that it is misleading.

It is notoriously difficult to spell out what it means to be 'in a position to know'. But it is perhaps enough to say that, just as it seems reasonable for Srinivasan to claim that Charles is in a position to know (indeed, knows) that the College master's counterarguments are misleading, it is reasonable to claim that both Nadja and Sarah are in a position to know that their interlocutors' counterarguments are misleading. The reason Nadja and Sarah do not know is just that they lack the requisite confidence that these counterarguments are misleading, not that they are in a worse epistemic position than Charles.

Third, I have supplied a diagnosis of *why* the Millian picture is appealing even though it is incorrect. I have argued that, if we view Mill's arguments as an exercise in full compliance theory, both their attractions and their deficiencies come into focus. Their attractions come into focus because, if we ignore the prevalence of persistent patterns of epistemic exclusion, it is plausible that engaging with challenges to our beliefs will secure certain epistemic benefits. But their deficiencies also come into focus because, once we pay attention to epistemic exclusion and oppression, it is no longer surprising why engaging with challenges to our beliefs will often not secure the adverted benefits, at least for those of us who are excluded from the means of knowledge production. Of course, Fantl could also supply a diagnosis of why the Millian picture is misleading. But that diagnosis would differ from mine. It might have to do with the fact that Mill relies on a false but tempting picture of justification, for example.

Finally, where Srinivasan and Fantl both offer some support for my argument in this chapter, in his 2019 book *Knowing our Limits* Nathan Ballantyne defends a view that might seem to cause problems for my argument in this chapter. Setting aside some of the details, Ballantyne's claim is that, often, we have defeaters for our beliefs, but lack defeaters for those defeaters. In other words, we often have evidence against our beliefs, or are aware that there could be evidence against them, yet lack any reasons to reject this counterevidence. When this is the case, Ballantyne thinks that the beliefs in question are unreasonable. Because he thinks we are often in this situation, Ballantyne endorses a widespread (though not global) form of scepticism about the reasonableness of many of our ordinary beliefs.[4]

To see how this might work in practice, consider an example. Imagine you believe a controversial proposition like 'global warming is caused by humans'. Ballantyne's claim is that, if you have a defeater for this proposition (e.g. you are aware that some people have arguments that global warming is not caused by humans) but lack a defeater for this defeater (you have no idea where those arguments go wrong), then your belief that global warming is caused by humans is unreasonable.

Extending this a little, you might argue that something similar applies to subjects in inhospitable environments like Laurie in §6.3, or subjects for whom the environment is inhospitable, like Nadja and Sarah in §6.4. Laurie, Nadja, and Sarah have true beliefs yet lack the ability to deal with counterarguments and counterevidence to those beliefs. If we apply Ballantyne's line of argument, we get

---

[4] Ballantyne doesn't claim that, when you have an undefeated defeater for a belief, the belief in question is unjustified (see 2019, p. 252, ff. 5). This is because he is happy to accept (at least for the sake of argument) the possibility that knowledge or unjustified belief can be unreasonable (Lasonen-Aarnio 2010). So there may be less of a tension between his view and mine than there initially appears to be.

the result that their beliefs are unreasonable. This goes naturally with the further claim that, when we need to deal with a challenge to our belief for us to be reasonable in holding on to the belief, we have an obligation to engage with that challenge. If this is right, then we can say that Laurie, Nadja, and Sarah were all under obligations to engage with challenges to their beliefs because, otherwise, their beliefs are unreasonable.

While Ballantyne has put his finger on an interesting sceptical problem (more on this in Chapter 8), he is too quick in drawing his sceptical conclusions. As Fantl argues in his review of Ballantyne's book, it may be that someone who is unable to pinpoint how exactly a piece of counterevidence is misleading (or where exactly a counterargument goes wrong) still has reasons for rejecting that piece of counter-evidence (Fantl 2020). We can distinguish between *targeted* and *holistic* (defeater) defeaters. You may lack a targeted defeater for arguments against your belief that global warming is caused by humans because you have no idea where those arguments go wrong (you are not a climate scientist!). But it could still be that your total body of evidence favours your belief, in which case you have a holistic defeater for these climate sceptical arguments.[5]

Applying this to Laurie, we could say that, even though she is unable to explain where exactly arguments against her beliefs go wrong, her total body of evidence still makes those beliefs reasonable. We might add that it could be that Laurie herself is not necessarily in a position to recognize that her total body of evidence favours her beliefs. But to require that she be able to recognize that it favours her beliefs would be to put a strong constraint on reasonable belief. Similarly, we could say, of Nadja and Sarah, that their evidence favours their readings of classroom and workplace gender dynamics over the readings of their interlocutors. Of course, they also do not recognize that their evidence favours their readings; in fact, they end up concluding that they are in the wrong, not their interlocutors. But the point is that it *would have been reasonable* for them to continue believing what they believed before interacting with their male colleagues.

6.7 Towards a Non-Ideal Theory

Let me return to the main line of argument. I have argued for a negative claim: we are not all under an obligation to engage with challenges to our beliefs. The

---

[5] This distinction is similar to a distinction Duncan Pritchard has drawn between discriminating and favouring evidence (see Pritchard 2010, 2012). Pritchard argues that traditional sceptical arguments might show that we lack discriminating evidence that we are not in certain sceptical scenarios (perhaps we lack evidence that discriminates between a scenario in which we are looking at a zebra and one in which we are looking at a cleverly disguised mule). But they do not show that we lack evidence that favours us not being in these sceptical scenarios (it may be that we have evidence that favours us being in the scenario where we are looking at a zebra, such as evidence that zoos do not typically deceive their customers like this).

flipside of this is that some of us are under such an obligation. Some of us are likely to derive epistemic benefits from engaging. But who is under this obligation and who is not? How can I tell if I am under it, whether in general or in a particular situation? In this section, I give a sketch—though only a sketch—of a positive account of our obligations as they pertain to challenges to our beliefs. My sketch is based on a discussion by Jennifer Lackey in her 2018 paper 'Silence and Objecting' of our obligations pertaining to testimony with which we disagree. On Lackey's view, whether you are under an 'obligation to object' depends on your social position. I will give a rough overview of Lackey's account, before indicating what a parallel account of our obligations pertaining to challenges to our beliefs would look like.

Lackey develops her view in response to a view defended by Sandy Goldberg in an unpublished paper (Goldberg manuscript). According to Goldberg, recipients of testimony have an obligation to object to testimony with which they disagree. The basic idea is that, if the recipient of testimony does not object, they risk misleading other participants in the conversation (or mere bystanders) about their view of the content of the testimony. Put crudely, the thought is that 'silence indicates assent'. Put a little less crudely, the thought is that, from the fact that the recipient of testimony does not indicate that they reject the content of the testimony, we are (defeasibly) entitled to infer that they accept it. Goldberg thinks that silence indicates assent because conversations are governed by Gricean cooperative norms and silent rejection is uncooperative. You therefore have a reason to think that silence indicates assent insofar as you have a reason to think you are having a (Gricean) conversation.

Lackey's basic objection to Goldberg is that he treats an ideal communicative exchange, where Gricean norms of cooperation are in play, and there is no asymmetry in power between speaker and hearer, as the normal situation. If we treat this as the normal situation, then actual exchanges in which such norms are not in play and there is an asymmetry in power must be viewed as departures from the norm. But, as Lackey points out, treating cooperation as the norm is problematic. It is not just that some conversations are not cooperative. Some people are more likely to find themselves in non-cooperative conversations than others. As Dotson has highlighted (recall §6.4), members of some social groups often find that their interlocutors are unable or unwilling to put in the work to try to understand what they have to say, or even to take them seriously. That cooperative conversations are the norm for some does not mean that they are the norm for all.

Lackey does not just critique Goldberg on the grounds that he works with an idealized picture of communication. She also develops her own 'non-ideal' account of a duty or obligation to object. She argues that whether you have an obligation to object to testimony you disagree with depends on your social position. So, for example, a White tenured male professor may have a duty to object when an assistant professor makes a racist remark while a Black female graduate student may not:

> [W]ith great power in a domain often comes greater authority, and thus an increased likelihood that one's testimony will have an effect. So, if we assume that the sexist remark in question is false and that one of our aims as epistemic agents is to promote the truth, then the white professor objecting to it might have more epistemic impact in producing true beliefs, both at the individual and collective level.   (2018, p. 92)

The White male professor's objection may stand a good chance of 'promoting the truth' by correcting any misconceptions that the remark may cause or solidify. But the Black graduate student would stand far less chance of achieving this result were she to object. Further, she is likely to face social costs if she speaks up. Because her objecting would have no tangible epistemic benefits and would have genuine costs for her, she is under no obligation to do so. Of course, if she were to speak up anyway, she would be doing something admirable, if perhaps futile. But actions can be admirable without being obligatory.

There are clear parallels between Lackey's critique of Goldberg and my argument in this chapter. Lackey argues that whether you have an obligation to object depends on whether you are likely to 'promote the truth' by doing so, but whether you are likely to promote the truth by speaking up depends in turn on your social identity and status. Similarly, I have argued that whether you have an obligation to respond to challenges to your beliefs depends on whether you are likely to secure epistemic benefits, whether for yourself or others, by doing so, but whether you are likely to secure these benefits depends in turn on your social identity and status. This is because whether you are likely to secure these benefits depends on whether you will be taken seriously when you speak and whether you will be taken seriously is influenced by social identity prejudices and stereotypes.

Building on this, compare Nadja and Sarah from §6.4 with two other agents, Stuart and Paul. Stuart and Paul have the exact same evidence and beliefs as Nadja and Sarah had before they spoke to their male colleagues. But, when Stuart and Paul speak to their male colleagues, their testimony is taken seriously—so seriously that, after further discussion, everyone comes round to Stuart's and Paul's views. As a result, all the male students make a concerted effort to behave better in discussions and, in Paul's case, an investigation is started into the behaviour of the senior male colleague. In these cases, it does seem more plausible that Stuart and Paul were under an obligation to respond to the initial challenges to their views. The reason it seems more plausible is precisely because their testimony was likely to be taken seriously and achieve tangible results.

Still, you might wonder how you are supposed to tell whether you are in Nadja and Sarah's situation, or in Stuart and Paul's. More generally, you might wonder how you are supposed to tell whether you are under an obligation to engage with challenges to your beliefs. In my view, it is a mistake to expect it to be a simple matter to figure out which obligations we are under, or indeed whether we are doing a decent job of fulfilling our obligations. These things are often obscure to us. Moreover, it is also a mistake to expect that the non-ideal epistemologist will be attracted to the idea that something can only be an obligation if we are able to recognize it as such, or if we are able to discern whether we are doing a decent job of fulfilling it. The dispute between ideal and non-ideal epistemology does not boil down to a dispute about whether our obligations (or whether we are fulfilling them) are transparent to us.

That said, my view is that figuring out whether you are under an obligation to engage with challenges to your beliefs is not really a matter of figuring out whether your epistemic position is particularly strong (as it would be on Fantl's view). It is a matter of figuring out whether you are likely to benefit (or whether others are likely to benefit) from you doing so. To be sure, this is not something that we are necessarily well placed to judge. But I see no reason to hold that we are rarely able to make reasonable judgements about these things. Part of learning how to navigate your social environment is learning which sorts of interactions are likely to yield positive results and which sorts are not worth your time and effort.

Let us take stock. I started this chapter with two arguments for the existence of a general obligation to engage with challenges to our beliefs. The first argument, which can be found in Mill, is that engaging with challenges to our beliefs secures several epistemic benefits. The second, which can be found in Mill but is developed in more detail by Cassam, is that it is only by engaging with challenges to our beliefs that we can earn the right to our beliefs. I then presented several cases that make trouble for both arguments. In these cases, it is hard to see why engaging with challenges will secure any epistemic benefits, and it is unclear why the agents in these cases would need to do so to earn the right to their beliefs.

Moreover, I provided a diagnosis of where Mill's and Cassam's arguments go wrong. They go wrong because they assume that other inquirers will fulfil their obligations and responsibilities and that the environment is epistemically hospitable. These are the hallmarks of full compliance theory, so we can say that the arguments go wrong because they are examples of ideal epistemology. I finished the chapter by comparing my argument in this chapter to other relevant literature and by offering a sketch of a non-ideal theory of our obligations pertaining to challenges. On this account, whether we are under this obligation depends on our social identity and status. The next chapter develops this thought further by considering a picture on which epistemic agency is itself thoroughly socially situated.

7

Liberatory Virtue and Vice Epistemology

In the previous chapter, I argued that whether you are under an obligation to engage with challenges to your beliefs depends on your social identity and status. More generally, I suggested that the nature and extent of our epistemic obligations and responsibilities depends on certain aspects of our social situation. While it is true that social interaction can have epistemic benefits, those benefits are not distributed equally. Whether individuals stand to benefit epistemically from social interaction depends on their social identities and situations. In a society where some of us face epistemic exclusion and oppression, it may be that those of us who are excluded from the means of knowledge production gain little or no epistemic benefit from certain forms of social interaction.

An epistemology that ignores all this is an ideal epistemology in the pejorative sense. The problem with it is not that it does not 'apply to the real world'. The problem is that, insofar as it applies to the real world, it is more likely to worsen than improve our epistemic situation because it leads inquirers to regard themselves as under obligations that they are not in fact under. In trying to meet these supposed obligations they will often end up making things worse rather than better. What we need is a different, non-ideal picture of our epistemic obligations and responsibilities. In this chapter, I develop a non-ideal picture using the theoretical framework of liberatory virtue and vice epistemology.[1]

The leading idea of liberatory virtue epistemology is that what it is to be an epistemically responsible agent and what is involved in manifesting the various intellectual virtues depends on aspects of social situation, such as social identity and roles. My main task in this chapter is to deal with a challenge for liberatory virtue epistemology. On the liberatory virtue epistemologist's picture, inquirers are deeply socially situated, and their character traits are shaped by social forces and influences largely outwith their control. The challenge is to make sense of the idea that we are still responsible for our intellectual virtues and vices, and the consequences thereof, even though they are shaped by factors outwith our control.

I will argue that the version of liberatory virtue epistemology developed by José Medina in his 2012 book *The Epistemology of Resistance* has the resources to deal with this challenge. As we will see, Medina has the resources to deal with it because

---

[1] In what follows I talk interchangeably of 'virtue epistemology' and 'vice epistemology'. This is because my focus in this chapter is José Medina, and he develops an integrated account of intellectual virtue and vice.

Non-Ideal Epistemology. Robin McKenna, Oxford University Press. © Robin McKenna 2023.
DOI: 10.1093/oso/9780192888822.003.0007

he urges us to rethink how we understand epistemic agency and responsibility. While he does not use these terms, Medina proposes a non-ideal picture of responsible epistemic agency. This chapter then, like the previous chapter, develops the second key aspect or face of non-ideal epistemology, which is a picture of epistemic agents as deeply socially situated.

Here is the plan. I start by introducing the basic framework of liberatory virtue epistemology (§7.1) and Medina's version of liberatory virtue epistemology (§7.2). The rest of the chapter is then devoted to raising the challenge for liberatory virtue epistemology (§7.3) and explaining why Medina has the resources to deal with it (§7.4). I finish the chapter by drawing out some connections with the argument of this book as a whole (§7.5).

7.1 Liberatory Virtue Epistemology

Let us start with what virtue epistemology is. It is an umbrella term for several different approaches to epistemology and it is surprisingly difficult to say what exactly unifies them. Some virtue epistemologists are primarily interested in the cognitive abilities or competences you must manifest to have knowledge (Greco 2010; Sosa 2007, 2009). For these virtue epistemologists, talk of 'intellectual virtues' is really talk of cognitive faculties such as perception. To have a 'faculty virtue' is, roughly, to have the capacity to reliably form true beliefs via the relevant faculty. These faculty virtues are interesting primarily because they are connected with knowledge. Indeed, the primary aim of 'virtue reliabilists' like Sosa is to utilize these faculty virtues in developing a theory of knowledge.

Other virtue epistemologists are more interested in identifying the cognitive traits required for intellectual flourishing (Baehr 2011; Roberts and Wood 2007; Zagzebski 1996). What these virtue epistemologists mean by 'intellectual virtue' is closer to what virtue theorists in ethics mean by 'moral virtue'. For these virtue epistemologists, intellectual virtues are understood as character traits. To have a 'character virtue' like open-mindedness is to be disposed to behave in certain ways—ways characteristic of open-mindedness. While some 'virtue responsibilists' (especially Zagzebski 1996) have tried to utilize these character virtues in developing a theory of knowledge, the more recent trend is to focus on the virtues themselves (see Baehr 2011 on 'autonomous virtue epistemology' or Roberts and Wood 2007 on regulative epistemology). Central questions for these virtue responsibilists include how to understand the various intellectual virtues, what is required to possess this-or-that intellectual virtue, and the value of the intellectual virtues themselves.

My focus in this chapter will be on this second, responsibilist, branch of virtue epistemology. But there is a further distinction that must be drawn within this branch. 'Conventional' or 'traditional' responsibilist virtue epistemologists assume that much the same traits make for intellectual virtue no matter how you are socially situated. They also do not consider the impact that your social situation has on the likelihood that you will develop a given virtue, or what the implications for virtue epistemology would be if it could so depend. Roberts and Wood's brand of virtue epistemology, which I discussed in Chapter 5, is a prime example of traditional (responsibilist) virtue epistemology. Their aim is to sketch a picture of the virtuous epistemic agent. This is an agent that we should all aspire to emulate, no matter the particulars of our social situation.

In her excellent overview of feminist and liberatory virtue epistemologies, Nancy Daukas argues that traditional virtue epistemologies like Roberts and Wood's go wrong precisely because they ignore social situatedness:

> Differences in social positioning matter to virtue epistemology because prevailing epistemic norms and practices impose different expectations on, and grant different opportunities to, those who enjoy different levels of social power. As a result, there are experiential, developmental, and normative differences between being a member of a group with more, or with less, power and privilege. Those differences matter to epistemic agency: an individual's epistemic capacities and dispositions develop to enable her to function from the position in which she finds herself. Individuals in different positions may develop different habits of attention, different styles of epistemic engagement, and different epistemic goals. They face different epistemic challenges and enjoy different epistemic opportunities.
>
> (2019, p. 380)

Daukas' claim is that what responsible or virtuous epistemic agency looks like depends on your social situation. This is very much of a piece with the argument of the previous chapter. In Daukas' terms, I argued that there are important experiential, developmental, and normative differences between inquirers who are persistently excluded from the means of knowledge production and inquirers who are not. These differences have consequences not just for the epistemic capacities and dispositions we should expect inquirers to develop but also for the capacities and dispositions we should view as required for responsible epistemic agency. Responsible epistemic agency for a privileged inquirer may include engaging with a wide range of challenges to their beliefs. Responsible epistemic agency for an oppressed inquirer may include engaging with a narrower—or different—range of challenges.

While Daukas is targeting traditional virtue epistemology, not ideal epistemology, her critique converges with my critique of ideal epistemology. It therefore makes sense to look at what Daukas thinks should replace traditional virtue epistemology as a way of developing my argument in the previous chapter. If Daukas' target is a form of ideal epistemology, it is natural to suppose that the alternative to it she suggests will be a form of non-ideal epistemology. Specifically, it will be a form of non-ideal virtue epistemology.

Daukas recommends that we replace traditional virtue epistemology with liberatory virtue epistemology. For Daukas, a liberatory virtue epistemology does two things. First, it recognizes that inquirers are socially situated, and that responsible inquirers must be sensitive to features of social situation, such as the power differentials that exist between inquirers. Second, it is guided by liberatory values in that its aim is to identify the epistemic practices and understandings that are complicit in and sustain oppression, and to replace them with practices and understandings that serve to combat oppression. In what follows I say a little more about each aspect of liberatory virtue epistemology.

What is involved in viewing inquirers as socially situated? Daukas tells us that traditional virtue epistemologists focus on 'virtues of self-reliance' like intellectual autonomy (recall Chapter 5) and intellectual courage. In contrast, liberatory virtue epistemologists focus either on relational virtues like empathetic caring (an ability to put yourself in another's shoes) or on reconceptualizing the traditional virtues in ways that foreground sensitivity to social situatedness. For instance, what Daukas calls 'liberatory intellectual autonomy' differs from traditional conceptions of intellectual autonomy (like those discussed in Chapter 5) in that it has little to do with self-reliance. The idea is more that the intellectually autonomous individual has the capacity and willingness to recognize problematic assumptions and commitments, including one's own problematic assumptions and commitments.

More generally, the 'traditional virtues' (the virtues of self-reliance) are character traits that enable those who possess them to function well within the world as we find it with all its injustices while 'liberatory virtues' are traits that enable those who possess them to recognize these injustices as injustices, as well as to combat them. Take what Daukas says about open-mindedness:

> [A]s conventionally understood, open-mindedness rarely involves the motivated critical acuity and courage through which to recognize that established norms are socially constructed, contingent, and laden with values, commitments, and patterns of behaviour that sustain conditions of injustice. Conventional open-mindedness doesn't destabilize the existing power structure or its representation of difference; at best, it encourages tolerance of differences as conventionally understood.   (2019, p. 387)

On the traditional picture, open-mindedness involves critically engaging with alternative perspectives. On the liberatory picture, it involves critically engaging with specific perspectives—the dominant ones, the ones that constitute existing power-structures. This is not the sort of trait likely to help anyone who possesses it to function well within the world as we find it. It is more likely to do the opposite. Someone who is aware of the ways in which established conventions, norms, and values are tied up in the perpetuation of injustice is not going to function well within the world as we find it. They are going to 'fight against' or resist it, not go along with how things are. This points to a further aspect of liberatory virtue epistemology, which is that the liberatory virtue epistemologist doesn't think there is necessarily a tight connection between being virtuous and flourishing, whether intellectual or otherwise (see Tessman 2005 on 'burdened virtues').

Turning to the second aspect, liberatory virtue epistemology is guided by liberatory values in the sense that it conceives of intellectual virtues as character traits that further liberatory political values. The virtuous epistemic agent is the agent with the ability to destabilize existing power structures due to their understanding of the unjust nature of these structures.

This might make it sound like, for the liberatory virtue epistemologist, intellectual virtues are defined by reference to politics rather than truth. This, in turn, might lead to the suspicion that liberatory and traditional virtue epistemologists are concerned with different things. The traditional virtue epistemologist is concerned to delineate character traits that serve epistemological ends. For the traditional virtue epistemologist, the virtuous epistemic agent is the agent who secures epistemic goods through exercising their virtues. In contrast, it might seem like the liberatory virtue epistemologist is concerned with which character traits would serve certain political ends, that is, the dismantling of unjust social structures.

This misunderstands the project of liberatory virtue epistemology. More generally, the objection that feminist and liberatory approaches to epistemology confuse politics with epistemology, or worse subvert truth and knowledge to power, misses the mark for reasons that have been thoroughly explained in the literature on feminist epistemology (Anderson 1995). The liberatory virtue epistemologist does not differ from the traditional virtue epistemologist in terms of the value they attach to truth and knowledge, or in whether they view the virtuous epistemic agent as the agent who secures (or tries to secure) epistemic goods. The difference has to do with which truths (and which goods) they attach value to.

It is a familiar point that the virtuous epistemic agent does not just seek truth (or knowledge). An agent who spent their life acquiring trivial truths (there are x grains of sand on this beach, there are y grains of sand on that beach) would not be intellectually virtuous. Rather, the virtuous epistemic agent seeks *significant* truth (Sosa 2003). While conventional virtue epistemologists differ amongst themselves as to how to determine significance, for the liberatory virtue epistemologist significance is determined by reference to liberatory goals. As Daukas puts it:

> [Liberatory virtue epistemology] promotes the cultivation of capacities and traits that successfully arrive at truths and understandings that are socially beneficial in that they empower epistemic agents to produce knowledge and understanding that is useful for dismantling and replacing conditions of oppression.
>
> (2019, p. 381)

The liberatory virtue epistemologist is concerned with delineating the traits that facilitate contact with *socially significant* truths. But socially significant truths are truths.

This completes my overview of liberatory virtue epistemology. I will now turn to a detailed examination of the best developed version of liberatory virtue epistemology available, which is that defended by José Medina in his *Epistemology of Resistance*.

7.2 Medina on Intellectual Virtue and Vice

My aim in this section is not to give an overview of Medina's entire liberatory virtue epistemology. It is rather to highlight the key claims that Medina makes and to anticipate some of his ideas about epistemic responsibility that will be important in the rest of this chapter. What follows draws on McGlynn (2019), which is an excellent overview of Medina's views of intellectual virtue and vice.

Medina develops two leading ideas in liberatory epistemology. The first idea, which he takes from feminist standpoint theory, is that your social situation shapes what you can and cannot know. Further, while socially marginalized or oppressed groups may be materially disadvantaged in many respects, they are epistemically advantaged in certain respects. While the ins and outs of the 'epistemic advantage thesis' is debated within standpoint theory, the idea is that it is, at the very least, easier for members of marginalized groups to attain certain sorts of knowledge than it is for members of privileged groups to attain those sorts of knowledge (Collins 1986; Harding 1991; Hartsock 1983. For a more recent discussion, see Toole 2019, 2022).

The second idea, which Medina takes from the epistemology of ignorance, in particular Charles Mills' work on 'white ignorance', is that there is a form of ignorance that is 'active' (Mills 1997, 2007). 'Active ignorance' occurs due to the activities of the ignorant subject rather than, as it were, 'by accident'. Active ignorance is important because, on Mills' view, it is complicit in creating and sustaining oppressive social structures. The thought is that oppressive systems require that those who benefit from them remain (at least to some degree) ignorant of how the system works, and perhaps even of the very existence of the system.

On Mills' view, the system whereby Whites dominate society and occupy almost all positions of political, economic, and legal power—the system of White supremacy—requires that those who benefit from it are ignorant of it, even to the extent that they refuse to acknowledge it exists. The idea is that it is hard to view yourself as the beneficiary of an oppressive system. One way of avoiding viewing yourself in this way is to remain ignorant of the fact that you benefit, or even of the fact that the system exists in the first place.

Medina develops both these ideas using the terminology of liberatory virtue epistemology. In Medina's hands, the idea that your social situation shapes what you can and cannot know is translated into the claim that your social situation shapes the intellectual virtues and vices you are likely to develop. Those in privileged social groups tend to develop a package of intellectual vices while those in oppressed groups tend to develop the correlating package of intellectual virtues:

The 'vices of the privileged':	The 'virtues of the oppressed':
Intellectual arrogance	Intellectual humility
Intellectual laziness	Diligence
Closed-mindedness	Open-mindedness

The privileged tend to be arrogant because they are less likely to be corrected when they make false claims. They tend to be lazy because they often have no incentive to figure things out for themselves. Moreover, they typically have an incentive to actively misunderstand the society around them and the ways in which they benefit from how it is structured. (This goes back to Mills' point that it is hard to view yourself as the beneficiary of an oppressive system.) Finally, they tend to be closed-minded because they are less likely to be open to anything that challenges their view of what society is like, or indeed of what they themselves are like.

In contrast, the oppressed tend to be intellectually humble because they are likely to be corrected when they make false claims. (They are also likely to be 'corrected' when they make true claims; this goes back to Dotson's work on epistemic exclusion and oppression, which I discussed in Chapter 6.) They tend to be intellectually diligent because they often have an incentive to figure out how the world works (their survival may depend on it). Finally, they tend to be open-minded because they are often forced to engage with alternative perspectives—and even to demonstrate respect for them.

You might think that these claims are too simplistic. Surely lots of privileged people will lack these vices, and many of those who are oppressed along some dimension(s) will lack the corresponding virtues. But the crucial point for my purposes is that it is plausible that your social position influences the character traits and virtues/vices you develop. This point stands even if Medina's account of how exactly this influence works is too simplistic. (For a more nuanced picture, see Tanesini 2021.)

Let me turn to how Medina develops the second idea. There are many ways of understanding active ignorance (see Martín 2021). You might understand active ignorance in psychological terms: it exists because those who benefit from the systems of oppression it upholds have a psychological interest in not having their self-understandings challenged. Or you might understand it in structural terms: it exists because the social structures that lead to oppression produce it. Medina proposes understanding active ignorance in psychological *and* structural terms:

> Active ignorance has deep psychological and sociopolitical roots: it is supported by psychological structures and social arrangements... Active ignorance is the kind of ignorance that is capable of protecting itself, with a whole battery of defense mechanisms (psychological and political) that can make individuals and groups insensitive to certain things, that is, numbed to certain phenomena and bodies of evidence and unable to learn in those domains.   (2012, pp. 57–8)[2]

As far as psychology is concerned, Medina views active ignorance as the result of a deeper psychological trait that he calls 'meta-insensitivity':

> For meta-blindness... or more generally and accurately, *meta-insensitivity*... is unlike any regular sort of blindness in the following respect: a blind person has no access to a whole range of situations but typically has an acute awareness that there is much more to the empirical world than what she or he can perceive. By contrast, the meta-blind ignoramus arrogantly assumes that there is nothing else to perceive beyond what she or he can see or hear... it is this persistent, stubborn denial that defines meta-blindness: *the inability to recognize and acknowledge one's limitations and inabilities.* This meta-blindness protects the first-order forms of blindness, which become particularly recalcitrant and resistant to change and improvement: cognitive deficits that are not even recognized are especially difficult to correct, and their correction will require transformations that the meta-blind subjects are ill-prepared to carry out by themselves.   (p. 76)

The meta-insensitive individual is ignorant of their own cognitive failings and limitations. This is a kind of second-order ignorance—an ignorance of your own ignorance. For Medina, meta-insensitivity tends to go along with the vices that are typical of the privileged. The privileged individual is not just often arrogant, lazy, and closed-minded. They are often ignorant of the fact that they are arrogant, lazy, and closed-minded. This is part of what Medina means when he says that active ignorance is accompanied by a 'battery' of defence mechanisms. Active ignorance is recalcitrant in part because it is a kind of ignorance that renders itself invisible, at least from the perspective of the actively ignorant.

Medina embeds these psychological structures within the social arrangements that are partly responsible for them. Following Fricker (2007), he emphasizes the influence of the 'social imaginary' on human cognition. Commenting on the Tom

---

[2] All the Medina quotes that follow are taken from Medina (2012).

Robinson trial that for Fricker is a central case of epistemic injustice, Medina says this:

> [I]n [the Tom Robinson trial] the *resistance to know* (to open one's mind to alternative possibilities and to ponder the available evidence fairly) comes from the *social imaginary* (or from limitations herein). It is the social imaginary that, in this case, breeds arrogance, laziness, and closed-mindedness. The social imaginary produces *active ignorance* by circulating distorted scripts about sexual desire according to which Negroes have a sexual agency out of control whereas white women lack sexual agency.   (p. 68)

The social imaginary produces active ignorance by circulating prejudices and stereotypes. Because these prejudices and stereotypes are widely held, lots of people have false beliefs due to their influence. But the social imaginary also produces the psychological structures that sustain this active ignorance. Medina continues:

> Those under the sway of this social imaginary—essentially all those who have been raised under the influence of these imaginings and the cultural representations they produced—are likely to develop epistemic habits that protect established cultural expectations and make them relatively blind and deaf to those things that seem to defy those expectations.   (p. 68)

We are all under the sway of the social imaginary and so influenced by various stereotypes and prejudices. But (this is the point Medina takes from standpoint theory) those who are oppressed or otherwise marginalized are more likely to *recognize* that this is the case than those who are privileged. In contrast, the privileged tend not only to have the vices of arrogance, laziness, and closed-mindedness but also to lack any awareness that they have these vices.

This might make it sound like, for Medina, the social arrangements are primary and the psychological structures secondary. There must be a sense in which this is indeed the case on Medina's picture. For Medina, the social arrangements produce the psychological structures. But, at least as far as the question of *responsibility* for intellectual vice, active ignorance, and epistemic justice is concerned, Medina does not want to abandon the idea that we are responsible for these things as individuals:

> We have already pointed out one of the crucial obstacles that factor into these listeners' inability to see things aright: the limitations of the social imaginary that colors their perception. So, in that sense, in order to address the injustice, we have to go well beyond the individuals involved in the exchange: we have to go to the *social roots* of the problem. But there is more. The failure to have the requisite

openness as a listener in testimonial exchanges involves the failure to establish the right personal connection with one's interlocutor. The problem is *both social and personal.*   (p. 80)

The idea is that, in situations of epistemic injustice, or other situations where active ignorance and the associated intellectual vices manifest themselves, there are both social and personal failings. The social failings are failings at the level of the social imaginary (we have a faulty 'interpretative toolkit'). While the personal failings are, in an important sense, the result of the social failings, that does not mean they are not still personal failings. This leads to the third and final aspect of Medina's view I want to highlight. Medina thinks that responsibility for active ignorance, epistemic injustice, and intellectual vice is *shared* rather than purely individual or collective:

> It would be a fatal mistake to assume that our diagnoses and treatments of hermeneutical injustices have to choose between the individual and the collective level…as I will argue below, without being able to operate at a crucial inter-mediate and hybrid level: the level of *shared responsibility.* It would be a mistake to think that in order to improve people's ability to make sense of things, to talk and hear about them, to participate in epistemic practices in more sensitive and fair ways, we need to choose either to meliorate individual's abilities or the community's resources.   (p. 82)

Medina's conception of responsibility is socialized but it still finds a place for individual responsibility. It is a conception on which individuals have responsibilities, but those responsibilities are tied to the communities they inhabit. But how exactly does this work? To get clear on what Medina is doing here, it will be helpful to take a step back and look at the problem he is trying to solve. The problem is how to make sense of individual responsibility on a picture where our characters are the products of social forces—of the 'social imaginary'.

7.3 Making Sense of Responsibility

In this section, I start by looking at some objections Robin Dillon has raised against feminist virtue ethics in her 2012 paper 'Critical Character Theory'. While Dillon does not discuss the prospects for feminist or liberatory virtue epistemology, it is not hard to see how what she says about feminist virtue ethics might carry over to liberatory virtue epistemology. As we will see, Dillon's point is that, from a feminist or liberatory perspective, we need to rethink responsibility. I will argue in the following section that Medina's view of responsibility recasts it in a way that should be acceptable from a liberatory perspective.

Dillon raises three problems for feminist virtue ethics. The first is that it neglects vices and other 'bad' character traits. Whether or not this is a problem with feminist virtue ethics, it is not a problem with contemporary liberatory virtue epistemology. As we have seen, Medina pays plenty of attention to intellectual vices. Alessandra Tanesini has also explored intellectual vice in detail, often from a (if perhaps implicit) liberatory perspective (see e.g. Tanesini 2016, 2018, 2021).

The second problem is that virtues and vices are usually understood as internal psychological features of individuals. As a result, talk of virtues and vices sits uneasily with the rejection of methodological individualism and the emphasis on social situatedness that is central to feminist or liberatory philosophy. While it is certainly true to say that traditional virtue epistemology sits uneasily with the methodological commitments of feminist philosophy, the whole point of feminist or liberatory virtue epistemology is that it is consistent with these commitments. As Daukas emphasizes, we can do virtue epistemology while recognizing that individuals are socially situated and acknowledging that their psychologies reflect their social situation. One of Medina's projects is to work through the details of all this. It is hard to see why we should think there is a deep tension here.

The third problem is more serious. Dillon argues that the language of virtue and vice smuggles in assumptions about blameworthiness and responsibility that are problematic once we recognize the role of social forces in shaping our characters:

> Just as the term "character" invites us, etymologically, to think of character as fixed, so the term "vice" and particular vice terms are typically understood to invite blaming individuals for having vices. It is not unreasonable to hear that invitation, for the term "vice" is related etymologically to terms for fault, failing, blameworthiness, crime, offense, violation, injury, and uselessness. But a revised understanding of character would call traditional activities of blaming into question, both as regards justification and as regards usefulness.   (2012, p. 105)

By 'a revised understanding of character', Dillon means an understanding on which our characters are not viewed as fixed or stable but as in large part the product of social forces. Dillon thinks that, once we understand character in that way, we also need to revise how we think about blameworthiness and responsibility:

> Nuanced examinations from a feminist critical perspective of particular "vices" could give us the means for talking about certain kinds of distortions of character without being focused on blaming and yet without dismissing the possibility that blaming may be both morally appropriate and useful, for fully acknowledging not only the ways in which sociopolitical conditions shape and partially

constitute character but also the ways in which individual agents, working "alone" and collectively, at the personal, intrapersonal, interpersonal, and social levels, have powers to take forward-looking responsibility for the kinds of persons that persons could become.   (2012, pp. 105–6)

Dillon is not arguing that we *cannot* make sense of the idea that we are responsible for our vices on a feminist or liberatory understanding of virtue and vice. Rather, she is arguing that, *if* we are going to make sense of responsibility on a feminist picture, we are going to need to (radically) rethink what it is to be responsible for your vices. We need a forward-looking account of responsibility that focuses more on the kinds of persons we could become than on the kinds of persons we are. The challenge then is to say what this new way of thinking about responsibility will look like.

To develop this point, it will be helpful to look at what has been said about responsibility for epistemic injustice and the intellectual vices that are tied up with it, such as prejudice. While Medina's developments of Miranda Fricker's account of epistemic injustice have been influential, it is surprisingly hard to find much discussion of his view of responsibility for epistemic injustice. On the other hand, there is a fair bit of discussion of Fricker's account of these matters. In the rest of the section, I will discuss some issues with Fricker's account that lend further support to Dillon's claim that we need to rethink responsibility.

Let us start with what Fricker says in *Epistemic Injustice*. Fricker holds that we *can* but *need not* be blameworthy for epistemic injustice. She contrasts two cases. In the first, which is taken from *The Talented Mr Ripley*, Greenleaf dismisses Marge's suspicions that another character in the story is a murderer on grounds that are clearly sexist ('there's female intuition, and then there are facts'). In the second, which is taken from Harper Lee's *To Kill a Mockingbird*, Tom Robinson is found guilty by the White jury even though compelling evidence of his innocence is presented to the jury. For Fricker, the jury find Robinson guilty because of the prejudiced beliefs they hold about him and his (lack of) credibility. In both these cases, then, someone is not believed by their audience because the audience harbours prejudices about them and the social identity group to which they belong. Fricker calls this *testimonial injustice*.

On Fricker's picture, those who perpetrate testimonial injustice do so because their credibility assessments are influenced by sub-conscious prejudices. This means that those who perpetrate testimonial injustice are usually unaware that they have done so. But this is not itself grounds for holding that you cannot be blamed for perpetrating testimonial injustice. You can be blamed for doing something even if you were unaware that you were doing it. Even if you are unaware that you have disregarded what someone has said because of a prejudice you hold about them, you may still be blameworthy for doing so. For example, you might have been able to access the reasons that spoke in favour of believing them.

This is precisely how Fricker proposes distinguishing between Greenleaf and the jury in the Robinson trial. Neither Greenleaf nor the jury were aware that their credibility assessments were influenced by prejudices. But, where Greenleaf did not have access to the reasons that would have been required to believe Marge, the jury did have access to the reasons required to believe Robinson. These reasons were presented clearly by the defence lawyer, Atticus Finch. Because they were presented clearly, the jurors should have recognized that there was a tension between their assessment of Robinson's credibility and the facts as presented in the trial. This in turn should have prompted some self-reflection on what led them to make this assessment, and ultimately to a re-assessment of Robinson's credibility. Because this did not happen, the jurors can be blamed for the testimonial injustice they perpetrated.

In his 2021 paper 'Responsibility for Testimonial Injustice', Adam Piovarchy argues that Fricker can't maintain this differential treatment of these two cases. More importantly, he also argues that the reasons Fricker gives for not viewing Greenleaf as blameworthy are equally good reasons for not viewing the jury as blameworthy. The result is that Fricker (at least Fricker 2007[3]) is committed to the view that we are typically not blameworthy for perpetrating testimonial justice.

Piovarchy's objection is that the reasons Fricker gives for thinking that Greenleaf was unable to access the reasons required to believe Marge—his sexist prejudices prevented him from recognizing them—apply equally to the jury in the Robinson trial. If the jury were not able to recognize Robinson as credible due to racist prejudice, then they were also unable to recognize the tension between their assessment of Robinson as a testifier and the plain facts of the case. As Piovarchy puts it:

> [T]he troubles with phrasing things in terms of "accessing reasons" is that many people cannot access reasons precisely due to the sort of person that they are. Since prejudice causes the jurors to not see Robinson as telling the truth, prejudice could also prevent them from noticing any conflict between the evidence and their assessments, or that any such conflict calls for a reflexive critical awareness.   (2021, p. 603)

If Fricker thinks that the obstacles the jury face to recognizing that Robinson is telling the truth are insurmountable, then the obstacles they face to recognizing the tension between their assessment of his credibility and the evidence presented in the trial are also insurmountable. If you also think, as many do, Fricker included, that you cannot be blamed for not doing something you were unable

---

[3] In her more recent work, Fricker defends an account of blame that may be able to sidestep Piovarchy's objections (e.g. Fricker 2016). I set this aside because my interest is in Medina, not Fricker.

to do, then it follows that the jury were not blameworthy for the testimonial injustice they perpetrated.

As Piovarchy notes, the natural response to this is to accept that neither Greenleaf nor the jury are necessarily blameworthy, but insist that they are both responsible nonetheless. We need to divorce responsibility from blameworthiness and delineate a kind of responsibility that does not require blameworthiness. This is exactly what we saw that Dillon calls on feminist theorists of virtue and vice to do. But the way in which Piovarchy tries to divorce responsibility from blameworthiness does not do what Dillon has argued needs to be done.

Piovarchy draws on a standard distinction from the literature on moral responsibility between two ways of thinking about responsibility. Following Gary Watson (1996), an agent is 'accountability responsible' for something (e.g. an action) if it is appropriate to make moral demands of them and to hold them accountable for failing to meet those demands. This sort of responsibility is closely tied to blame. If it is appropriate to demand that someone do something, yet they do not do it, then they will be blameworthy for not doing it absent a good excuse or some other mitigating factor.

On the other hand, an agent is attributability responsible for something (e.g. an action, a character trait) if it reveals something about 'who they are'. The thought is that, while some actions do not reveal much about our 'deep characters' (they do not say much about 'who we are'), other actions do. It is these actions that we are attributability responsible for. This sort of responsibility is not closely tied to blame because we do not necessarily get to choose who we are. Still, what we do reveals who we are. If we have 'bad' qualities (e.g. if we are prejudiced) we can be held responsible for what we do because of our bad qualities.

Piovarchy has argued that, at least on Fricker's understanding of the cases, neither Greenleaf nor the jurors in the Robinson trial are accountability responsible for perpetrating testimonial justice. Because they are unable to prevent their prejudices from interfering with their credibility assessments, it would be inappropriate to demand them to do so or blame them for not doing so. But Piovarchy thinks it is plausible that both Greenleaf and the jurors in the Robinson trial are attributability responsible for perpetrating testimonial justice. Their actions (the conclusions they reached about Robinson's guilt and the credibility assessments they made in reaching those conclusions) reflect the fact that they are deeply prejudiced agents. This is so even if they cannot help but be prejudiced.

Is this a satisfactory account of responsibility for testimonial injustice? Would an extended version of it—a version encompassing other forms of epistemic injustice and active ignorance—be satisfactory? It is fine as far as it goes. But it is worth going back to the problems Dillon raised for feminist theorizing about virtue and vice to see why it is not exactly what the liberatory virtue epistemologist is looking for.

First, the idea that we are responsible for testimonial injustice in the sense that perpetrating it reveals something about our deep characters is plausible enough.

But it is not entirely clear how to extend this to other forms of epistemic injustice or active ignorance. The basic idea behind epistemologies of ignorance like that developed by Mills is that certain kinds of ignorance are endemic for complex historical, political, and social reasons. The sense in which my ignorance of, say, the ins and outs of the transatlantic slave trade and the role that port cities in the UK like Glasgow and Liverpool played in it, reveals something about my deep character is attenuated at best. It is part of the explanation why I am ignorant of these things, but it is a small part. The more important part of the explanation will focus on things like the British education system (Bain 2018) or 'collective amnesia' (Tanesini 2018).

Second, Dillon suggests that we need a 'forward-looking' way of thinking about responsibility—a way of thinking about responsibility focused on the 'kinds of persons that persons could become' and the kinds of things that persons might do. Both accountability and attributability responsibility are backward-looking ways of thinking about responsibility. They are focused on the kinds of persons we are and the kinds of demands that can be appropriately made on us given the kinds of persons we are. A complete account of responsibility—at least, an account that is complete by the lights of feminist or liberatory virtue epistemology—would incorporate both forward- and backward-looking elements. But the sort of account Piovarchy suggests only incorporates a backward-looking element.

In summary, the challenge for the liberatory virtue epistemologist is to not just make sense of responsibility for epistemic injustice but more generally to make sense of responsibility for active ignorance and the intellectual vices tied up with it. Moreover, the challenge is to do so in a way that is forward-looking as well as backward-looking. In the next section, I will argue that Medina's account of responsibility can do all of these things.

7.4 Medina on Epistemic Responsibility

In this section, I argue that Medina has a sophisticated account of responsibility that can avoid the problems Dillon raises for feminist theorizing about virtue and vice. Doing this will require sorting through the many and varied remarks Medina makes about responsibility for active ignorance, epistemic injustice, and intellectual vice in *Epistemology of Resistance*. As we will see, there are some strands of his thinking that run into problems like those raised by Piovarchy for Fricker. But there are also some strands that go in the direction demanded by Dillon.

We can start with the strands that run into problems. While Medina's analysis of the Robinson case is different from Fricker's, he gives a similar explanation of why the jury are responsible for the testimonial injustice they perpetrate. For Medina, the jury regard Robinson as lacking credibility because their interpretative tools are faulty. During the trial it becomes clear that Robinson was trying to help Mayella Ewing (the woman he is accused of sexually assaulting). But the jury cannot understand how a Black man could be in the position to help a White woman. As a result, they are incapable of taking the possibility that he is telling the truth seriously. Medina's thought is that their refusal to believe Robinson is underpinned by their inability to understand him and they are responsible for their inability to understand him:

> Hermeneutical gaps are performatively invoked and recirculated—*reenacted*, we could say—in the speech acts of daily life. And we have to take responsibility for how our communicative agency relates to the blind spots of our social practices ... We have to evaluate whether our communicative actions and inter-actions are contributing to interrogate and expand hermeneutical sensibilities or not.   (2012, p. 111)[4]

The jury needed to take responsibility for the 'blind spots' in their 'hermeneutical sensibilities' which rendered them unable to understand Robinson. Of course, it would be asking a lot to expect them to eradicate these blind spots overnight (or during the trial). But Medina does not think it would be asking too much to expect them to try to do so:

> Even if hearers cannot be expected to be able to suddenly develop a *complete* openness with respect to something they have been trained not to hear or to hear only deficiently, they can be blamed for not even trying in the least to interrogate their interpretative habits and to make an effort to put themselves in the shoes of the speaker and consider what she could possibly be trying to convey.   (p. 112)

Because they did not make any effort at all, they can be held responsible for the epistemic injustice they perpetrated.

Medina's basic move here is like Fricker's. Rather than holding that the jury is directly responsible for the prejudices that led them to perpetrate an epistemic injustice, he tries to locate something else for which the jury are responsible. For Fricker, the jury are responsible for their failure to recognize a tension between their assessment of Robinson's credibility and the plain facts of the case. For Medina, they are responsible for their failure to interrogate their interpretative habits and put themselves in Robinson's position.

As Piovarchy points out, the problem for Fricker is that it is hard to see how the jury could be responsible for their failure to recognize this tension if they are not also responsible for their more basic failure to recognize Robinson's credibility. The problem for Medina is that it is equally hard to see how the jury could be

---

[4] Again, all the references that follow are to Medina (2012).

responsible for their failure to interrogate their interpretative habits if they are not also responsible for their more basic failure to understand Robinson. The factors that explain why the jury do not interrogate their interpretative habits or empathize with Robinson also explain why they do not understand him (i.e. their deep-seated prejudices).

Another strand of Medina's thinking that runs into problems is his idea that epistemically responsible agency requires kinds of knowledge which the actively ignorant are likely to lack. It is a familiar thought that responsible agency requires some form of self-knowledge. If you are entirely 'in the dark' about your intentional states (what you believe, what you desire), it is hard to see how you could be regarded as a responsible agent at all. An agent who had no idea what they believed or wanted would be entirely unable to account for why they have done what they have done, and many hold that at least some ability to justify your actions is required for responsible agency (Bilgrami 2006). Medina agrees that at least some form of self-knowledge is required for responsible agency. But he also thinks that other forms of knowledge are required for responsible agency:

> *Some* self-knowledge is indeed required by rationality and responsible agency. But the same is true of other types of knowledge, such as knowledge of the world and knowledge of our fellows and their minds: they are also required by rationality and responsible agency; and no one can be considered a rational and responsible agent unless she has some *minimal* knowledge about the empirical and the social world, that is, *some* correct beliefs about her surroundings … to be a responsible agent is to be a *minimally* knowledgeable subject. Responsibility and epistemic competence are bound up with each other: there is no responsibility unless there is minimal knowledge about self, others, and the world.   (p. 127)

Medina's thought is that, if it plausible that responsible agency requires minimal self-knowledge, then it is also plausible that it requires minimal knowledge of the social world (knowledge about others) and the external world (knowledge about objective facts). He also thinks that these conditions for responsible agency are not always met:

> The threefold presumption of epistemic authority can be maintained only in the *absence of systematic distortions and of cultivated forms of blindness and ignorance*, that is, under conditions of *epistemic justice*. It is only when there are no pervasive epistemic injustices, no systematic roadblocks to the development of knowledge, that responsible agents should be expected to be minimally knowledgeable about themselves, their peers, and the world.   (p. 128)

We cannot assume that the actively ignorant, subjects who have the vices of the privileged, and the meta-insensitivity that typically goes with these vices, will meet the requirements for responsible agency. As we saw earlier, someone with these vices is typically ignorant of facts about themselves (about the vices they have), about others (about their credibility), and even of objective facts. As Medina puts it:

> My contention is that when an individual is epistemically irresponsible with respect to others, it is very often the case that he is also epistemically irresponsible with respect to himself, because his social ignorance also involves self-ignorance. ignorance about his own positionality and relationality with respect to those ignored others, and quite possibly also ignorance about certain aspects of himself that he is unable to recognize—for example, he may be unable to see how the particular configuration of his religious identity is built around the intolerance and exclusion of other religious identities.   (p. 143)

I am inclined to agree that, if some self-knowledge is required for responsible agency, then so is some interpersonal knowledge, and some knowledge about objective facts. But my present concern is whether this furnishes an argument that the actively ignorant are responsible for their ignorance. In this passage Medina elides the important distinction between being *epistemically irresponsible* and *lacking the capacity for epistemically responsible agency*. Medina's view is that, if you lack certain kinds of knowledge, then you cannot be regarded as an epistemically responsible agent. But this is not because you are epistemically irresponsible. It is because you are not even 'in the market' for being an epistemically responsible agent. If this is right, then far from furnishing an argument that the actively ignorant are responsible for their ignorance, Medina has really given us an argument that they are not even in the market for being responsible agents. Perhaps Medina is happy with this result, but it is a non-starter from the point of view of making sense of responsibility.

So far, I have considered strands in Medina's thinking that run into problems. I now want to turn to a final strand in Medina's thinking that is a lot more promising. In the second half of chapter 4 of *Epistemology of Resistance*, Medina develops a picture of responsibility that emphasizes two key themes. First, your responsibilities depend on your social position and role. Some social roles come with heightened responsibilities. We often expect more of public officials, educators, and parents not because those who occupy these roles necessarily have insights or knowledge that others lack but simply because they occupy these roles. As Medina puts it:

> [B]esides experts in this traditional and academic sense, there are also more ordinary subjects who bear special responsibilities in the production of ignorance through a social division of cognitive laziness, namely, those who are in a position to educate and are charged with the task of being vigilant about

epistemic lacunas, distortions, and cognitive deficiencies. In this respect, given their participation in formative processes, parents and teachers carry particularly demanding epistemic burdens in the social division of both cognitive labor and cognitive laziness. Parents and teachers can be both inducers and inhibitors of intellectual curiosity, facilitators and blockers of critical skills and transformative processes of learning. Given their roles and positions in the community and the special epistemic obligations they have undertaken or been assigned, they deserve special credit for the epistemic successes and failures in the cognitive development of individuals under their care.   (p. 147)

We are familiar with the idea of a social division of cognitive labour. We cannot figure everything out for ourselves because we lack both the time and the ability. As a society, we make decisions about who is to focus on what. Medical researchers develop treatments and vaccines. Engineers figure out how to build things. The flipside is that, if those who are assigned a role in the division of cognitive labour do not carry out that role properly, there is nobody else who is going to do it.

If we apply the idea of a social division of cognitive labour to active ignorance and epistemic injustice, we get the following result. There must be groups whose job it is to recognize and eradicate active ignorance and identify cases of epistemic injustice. If these groups do not carry out this role properly, pernicious forms of active ignorance will remain, and epistemic injustice will go unrecognized. Medina's suggestion is that parents and educators have special responsibilities in this respect. Their job is not just to impart knowledge but also abilities, skills, and—hopefully—the virtues needed to recognize active ignorance in yourself and avoid perpetrating epistemic injustices.

Importantly, those who occupy these roles have these jobs not because they are able to do these things but because they inhabit these social roles. Of course, you would hope that someone who occupies these roles is able to do these things. But the point is that you have these responsibilities in virtue of inhabiting the role, not in virtue of some special skill or talent you possess. Compare: you would hope that a research scientist can do the things expected of a research scientist. But they have the responsibilities typical of a research scientist in virtue of occupying this role not because they are able to do the things expected of them.

Second, this might make it sound like, on Medina's picture, some people— those who are given the work of 'being vigilant about epistemic lacunas, distortions, and cognitive deficiencies'—have all the responsibility while the rest of us are off the hook, so to speak. But this is not his view:

[T]he developing individuals as well as others more indirectly related to them— including the entire social body—should also be credited and held responsible for the epistemic successes and failures in the social production of knowledge and ignorance. *Note that my thesis of the heightened responsibility of some, which derives from the theses of social division of cognitive labor and laziness, does not exculpate other individuals and groups, or the community as a whole.*

(p. 147, my emphasis)

To say that some people have heightened responsibilities when it comes to active ignorance, epistemic injustice, and the like is not to say that others have no responsibilities. Medina's picture, then, is that we all have responsibilities as individuals (though these responsibilities may vary) and as a collective to combat active ignorance and epistemic injustice. This is what Medina means by shared responsibility. We are responsible as individuals *and* as a collective.

To see how this picture works, we can look at the analogy Medina draws with Iris Marion Young's 'social connection model' of responsibility (Young 2006, 2011). On Young's picture, everyone who participates in unjust social structures—and this is almost everyone—has at least some responsibilities to remedy the injustices these structures perpetrate. What these responsibilities are will depend on their social identities and roles. As in Medina's discussion of our responsibilities as inquirers, some individuals may have special responsibilities here (e.g. public officials). But they have these responsibilities not because they had a role in bringing these injustices about (though they may have had a role) or because they are to blame for their role (though they may in fact be to blame) but in virtue of their social positions and roles—their positions within the social structure.

Applying Young's model to active ignorance and epistemic injustice, we can say that, given (almost) everyone participates in social structures that produce active ignorance and perpetrate epistemic injustice, we all have at least some responsibilities for remedying active ignorance and epistemic injustice. Some of us may have heightened (or simply different) responsibilities (e.g. educators or parents), but nobody is 'off the hook'. We all have a responsibility to interrogate our testimonial and hermeneutical sensibilities, our prejudices and biases, and our character traits. We also have a responsibility to listen—most of all to people who are not like us. Crucially, we have these responsibilities simply because we participate in these structures, not because we are necessarily intimately involved in perpetrating epistemic injustices (though we may be) or because we are necessarily to blame for our role in bringing them about (though it may be that we sometimes are).

I have only sketched the broad outlines of Medina's account of our responsibilities as inquirers. A good deal more would need to be said about how exactly responsibilities are assigned to individuals via their social identities and roles. But I hope what I have said suffices for my main purpose, which is to show that Medina's account of responsibility promises to do what Dillon said a feminist or liberatory account of responsibility must do.

On Medina's picture, we can go along with Piovarchy's suggestion that we think of responsibility for epistemic injustice in terms of attributability rather than accountability. This gives us a lot of what we want from a backward-looking conception of responsibility: it provides a sense in which those who perpetrate epistemic injustice are responsible for it in virtue of being the kinds of people they are. (Whether it gives us everything we want depends on whether it can be extended to deal with active ignorance and the like.)

But we can say a lot more than this. We can combine this with a forward-looking conception of responsibility on which we have various responsibilities not in virtue of being the kinds of people we are but because of the kinds of people we might become. We have responsibilities with respect to our intellectual character traits. These responsibilities include reflecting on them, improving them, and mitigating the consequences of our bad traits. We also have responsibilities when it comes to active ignorance, epistemic injustice, and the like. These responsibilities include identifying pernicious forms of active ignorance and persistent patterns of epistemic injustice. They also include trying to rectify them. (If we occupy certain social roles, our responsibilities in these respects are particularly weighty.) But these are not the sorts of responsibilities that are intimately tied to blame and blameworthiness or that we have because of who we are 'deep down'. We have them not because we are necessarily good at doing what they demand of us or because of the sorts of people we are but because we need to do these things to remedy existing injustices and make the world a better place.

7.5 Ideal Theory and Epistemic Responsibility

In this chapter, I have argued that liberatory virtue and vice epistemology is a promising way of developing the idea that our epistemic obligations and responsibilities depend on our social identities and positions. I have also argued that what many critics—including feminist critics, like Dillon—take to be a central problem for liberatory virtue epistemology can be dealt with, at least if we follow Medina's lead and develop a socially situated account of epistemic agency and responsibility. On a socially situated approach like Medina's, we have responsibilities *in virtue of* our social situatedness—our connections to others, our social identities, our social roles. Far from constituting a problem for the idea that we are responsible for our intellectual vices and the consequences thereof, our social situatedness is the ground on which our responsibilities are built.

The approach to epistemic responsibility I have outlined in this chapter dovetails nicely with some recent work on moral responsibility, the gist of which is that we need to view our social identities and positions as the grounds of our responsibilities to each other (Vargas 2018; Zheng 2018, 2021. See also the essays in Hutchison, Mackenzie, and Oshana 2018). It also fits nicely within a non-ideal epistemological framework. I want to finish this chapter by making the ways in which Medina's account of responsibility fits within a non-ideal framework a little more explicit. There are three points I want to highlight.

First, I have contrasted different ways of thinking about responsibility for intellectual vice and for the injustices and forms of ignorance that intellectual vices are tied up with. Some of these ways of thinking about responsibility are, in an obvious sense, idealized. The idea that we might be responsible for our vices in the sense that we can exercise control over them runs into problems once you recognize the extent to which multifarious social forces not only shape our character traits but also limit our ability to change or even recognize those traits. Someone who is prejudiced because their society is prejudiced is as little able to do something about their prejudices as they are to avoid letting their prejudices inform their behaviour in the first place. The problems with thinking about responsibility for intellectual vice in these terms become clear once it is recognized that a picture of human agency as not constrained by multifarious social forces and influences is very much an idealized picture.

There are many ways in which you might try to improve on this idealized picture. The idea that we can be responsible for our vices in the sense that they reflect 'who we are' is promising because it allows that we may be who we are due to the influence of social forces. Still, thinking of responsibility in these terms does not so much socially situate responsibility as try to carve out an individualistic way of thinking about responsibility given the realization that we are socially situated. Medina (and Young) goes further and socially situates responsibility. On this picture, our social connections to others are the grounds of epistemic agency and responsibility. The result is a picture on which we have responsibilities because of our social positions.

The second point I want to highlight is more methodological. The difficulties the liberatory virtue epistemologist faces in reconciling their emphasis on social situatedness with the terminology of virtues and vices illustrates a more general challenge for non-ideal epistemology. As I noted in Chapter 2, the distinction between ideal and non-ideal epistemology is best viewed as a distinction between two contrasting tendencies. Where the ideal epistemologist is happy with certain idealizations, whether about human cognition, social interaction, or social insti-tutions, the non-ideal epistemologist wants to question these idealizations. Are they harmless, as those who make the idealizations want us to suppose, or do they distort our philosophical theorizing?

It is not possible to eschew all idealizations in philosophical theorizing. This means that the non-ideal epistemologist faces tricky questions about where we should idealize. We can view critics of liberatory virtue epistemology—especially feminist critics like Dillon—as arguing that the liberatory virtue epistemologist relies on idealizations (about character and character traits) that we would do well to eschew. I have argued that this objection fails, at least if we want to apply it to

Medina's version of liberatory virtue epistemology. But the general point that the non-ideal epistemologist has work to do in figuring out how non-ideal they want their epistemology to be is a good one, and important to keep in mind if you want to do what this book urges and adopt the non-ideal approach.

The third point I want to highlight concerns a misunderstanding of non-ideal epistemology and the difference between it and ideal epistemology. It is tempting to think of the debate between ideal and non-ideal epistemology as a debate about the extent to which our epistemic norms and ideals should be easy to follow and attain. On one side stands the ideal epistemologist, who wants to urge the importance of strict norms and lofty ideals. On the other stands the non-ideal epistemologist, who wants to say that it is all very well insisting on strict norms and lofty ideals, but we can hardly expect actual humans to live up to them.

This way of understanding the debate is very much at odds with how I have characterized the distinction between ideal and non-ideal epistemology. It also does not fit well with the objections I have levelled at ideal epistemology. My objection has typically been that the norms and ideals proposed by the ideal epistemologist run the risk of worsening our epistemic situation, not that they are unattainable.

Medina's account of our epistemic obligations and responsibilities is interesting in this respect because it is, in many ways, very demanding. Medina departs from the virtue epistemological tradition in several ways, but one way in which he is aligned with the tradition is that, on his picture, being virtuous is hard. Being virtuous requires regularly interrogating and reflecting on your assumptions, biases, beliefs, interpretive habits, and prejudices. It requires empathizing with others, particularly with people whose life and life experience is quite different from your own. Doing these things is not easy and Medina is under no illusions about our chances of success.

This might prompt a worry: is Medina open to the objection I have levelled at the ideal epistemologist, which is that trying to do the things he thinks we should do will worsen, not improve, our epistemic situation? The first thing to say is that, if he is, then that needs to be shown. Someone who wants to claim that trying to follow a particular norm or attain a particular goal will worsen our epistemic situation needs to provide evidence that this is so. This is how I proceeded in Chapters 3 (on Goldman), 5 (on Carter and Roberts and Wood), and 6 (on Mill and Cassam). If it turns out that Medina is also open to this objection, this is a reason to question many of his central claims. But it is a reason that comes from a non-ideal perspective and so provides no grounds on which to reject non-ideal epistemology.

The second thing to say is that Medina has some resources when it comes to this objection that are not necessarily available to the ideal epistemologist. For Medina, epistemic responsibility is shared. While we have a responsibility to engage in critical self-reflection, this is not just a responsibility we have as individuals but also a responsibility we have as a collective. It may well be that, if we leave individuals to their own devices, they are quite bad at critical self-reflection (the psychological literature I have cited in Chapters 3, 4, and 5 suggests this is true). But what about groups of individuals, particularly diverse groups of individuals? Do we have equally good reasons to think diverse groups are bad at critical self-reflection? It is, at the very least, not obvious why reasons for scepticism about the prospects of individual self-improvement carry over to the prospects of collective self-improvement.

Let me conclude. This and the previous chapter have illustrated the second key aspect or face of non-ideal epistemology, which is a view of epistemic agents as deeply socially situated. While a view of epistemic agents as socially situated is common in social epistemology, what differentiates the non-ideal perspective on epistemic agency and responsibility from the more ideal perspective is that, on the non-ideal perspective, our social situations—especially our social identities and roles—are the grounds of our obligations and responsibilities rather than potential constraints on them.

In the concluding chapter, I take up a question which might seem quite different from the issues I have been concerned with in previous chapters. The question is: does the psychological literature on motivated reasoning I have cited in earlier chapters provide the basis for an argument that our beliefs about political and politically contentious scientific issues are often unjustified? I argue that—with a few caveats—it does. But, as I also argue, this serves to illustrate the point I have been making in this chapter. Non-ideal epistemology can be very demanding—demanding enough that you can argue for a distinctive form of scepticism from a non-ideal perspective.

8

Scepticism Motivated

In this chapter, I take a closer look at motivated reasoning. My primary aim is to address the import of motivated reasoning for the justificatory status of beliefs that are impacted by it. My central question will be whether such beliefs are justified. There is some initial reason to think that beliefs that are affected by motivated reasoning are not justified. The 'motivated reasoning paradigm' says that our assessments of arguments and evidence are partly dictated by our wants, desires, and preferences (Lord, Ross, and Lepper 1979; Kahan 2016a; Molden and Higgins 2012; Taber and Lodge 2006). If this is so, how could these assessments, or the beliefs we form based on them, be justified? My aim in this chapter is to argue that this line of argument is more than just initially plausible. The prevalence of motivated reasoning in some domains (e.g. political cognition) should lead us to be far less confident in our beliefs in these domains than we usually are.

There are several reasons why the question of the epistemic status of beliefs affected by motivated reasoning is worth addressing. It is clearly politically timely. Yet, until recently, it has been ignored by those working in mainstream epistemology (for exceptions, see Avnur and Scott-Kakures 2015; Huemer 2016; Kelly 2008).

More importantly for my purposes, this chapter serves to underpin a point I have already made in several places. The non-ideal epistemologist's criticism of ideal epistemology is not that it traffics in norms that are difficult to follow or ideals that are hard to attain. Their criticism is that it often traffics in norms and ideals that we would do better not to try and follow or attain. In this chapter, I buttress these points by showing that there is a kind of scepticism that can be supported by considerations that are typical of non-ideal epistemology. Even if it turns out that this kind of scepticism should be rejected, the non-ideal epistemologist should not reject it simply on the grounds that it sets the bar for justified belief too high. The mere fact that a bar is hard to reach is not a reason to lower it.

Here is the plan. I start by recapping the empirical literature on politically motivated reasoning (§8.1). This serves as the basis for the sceptical arguments I develop in the following sections (§§8.2–8.4). I present two arguments, one based on the claim that beliefs formed via politically motivated reasoning are unreliable, and another based on the claim that beliefs formed through politically motivated reasoning are not based on whatever evidence the believer might have for them in the right sort of way to be justified. Finally, I discuss a response to these arguments, which is that we should lower the bar for justified belief. I explain why this response should not appeal to the non-ideal epistemologist (§8.5).

Non-Ideal Epistemology. Robin McKenna, Oxford University Press. © Robin McKenna 2023.
DOI: 10.1093/oso/9780192888822.003.0008

8.1 Even More on Motivated Reasoning

While I have discussed motivated reasoning in earlier chapters, my argument in this chapter requires me to do into more detail than I have previously. Here, as elsewhere in the book, I will largely focus on politically motivated reasoning, and in particular on work by Dan Kahan and collaborators (Kahan 2013, 2014, 2016a; Kahan, Jenkins-Smith, and Braman 2011). For Kahan, the motivation or goal that is served by politically motivated reasoning is *identity protection*. The goal is to form beliefs that protect and maintain our status within a group that defines our identity and whose members are united by a shared set of values. We can therefore expect people to engage in politically motivated reasoning when they think about issues that have become entangled with questions of cultural and social identity. As Kahan puts it:

> When positions on some risk or other policy relevant fact have come to assume a widely recognized social meaning as a marker of membership within identity-defining groups, members of those groups can be expected to conform their assessment of all manner of information—from persuasive advocacy to reports of expert opinion; from empirical data to their own brute sense impressions—to the position associated with the respective groups.   (2016a, p. 1)

While untangling the various cultural and social identities that drive politically motivated reasoning would be a worthwhile task, Kahan, following others in the literature, tends to talk in more narrowly political terms, especially when it comes to interpreting data and drawing more general morals. In this chapter, I will follow this practice and use the crude labels 'liberal' and 'conservative' to pick out two quite different political groups, the members of which disagree on a wide range of political and politically relevant scientific issues. As before, I use both labels in the way they are typically used in discourse about contemporary US politics.

Let me summarize three key results from the literature on politically motivated reasoning. These results have been discussed in earlier chapters, but for the purposes of making my argument it will be helpful to collect all the results I need in one place.

First, the empirical evidence does not suggest that politically motivated reasoning is less prevalent in 'knowledgeable subjects'. In fact, it suggests the opposite. In Chapter 5, I discussed an important study from Taber and Lodge (2006), which found that, the more knowledgeable a subject was, the more likely they were to conform their assessment of the quality of the arguments and information they consider to their political beliefs. Taber and Lodge's explanation of this result is that it is due to two widespread cognitive biases. We tend to spend more time and resources finding flaws with arguments that challenge our deeply held political beliefs than we do on finding flaws with arguments that support these beliefs.

Moreover, when we are free to choose which information to expose ourselves to, we tend to seek out arguments that will confirm our deeply held political beliefs rather than arguments that will challenge them. The more knowledgeable you are—the more information you have at your disposal—the better you will be at finding flaws in arguments with conclusions you do not like, and the better you will be at seeking out information that confirms rather than challenges your existing beliefs.

Second, our political beliefs impact on our evaluations of who the experts are. Recall the study from Kahan, Jenkins-Smith, and Braman (2011) which found that our assessments of the level of expertise of (fictional) scientists correlate with how well their positions on topics such as global warming and nuclear waste disposal conform to our political beliefs. Put simply, we regard those who take positions we disagree with as having less expertise than those who take positions with which we agree for the simple reason that their views differ from our own.

This suggests a reason that the simple advice to 'follow the experts' is not particularly useful as a persuasive strategy. While we might want laypersons to apply neutral conditions for identifying experts about topics like global warming, laypersons in fact form impressions about who the experts are that are influenced by their prior beliefs. As I discussed in Chapters 3 and 4, it is for this reason that telling people to look at what the experts say is often not a successful strategy for defusing debates that have become politically contentious. Different sides in the debate will differ on who the experts are, so we cannot just tell them to turn to the experts to decide the issue in question.

Third, most of us engage in politically motivated reasoning, at least some of the time. Kahan and his collaborators look at the impact of our political beliefs on our views about topics including global warming, the safety of 'burying' nuclear waste underground, and 'concealed carry' laws, among others. Kahan, Jenkins-Smith, and Braman (2011)[1] report the following findings:

- 78 per cent of liberals think most scientists agree that global temperatures are rising, whereas only 19 per cent of conservatives think most scientists agree. (56 per cent of conservatives think most scientists are divided and 25 per cent think most disagree.)
- 68 per cent of liberals think most scientists agree that humans are causing global warming, whereas only 12 per cent of conservatives think most scientists agree. (55 per cent of conservatives think most scientists are divided and 32 per cent think most disagree.)

---

[1] Kahan, Jenkins-Smith, and Braman split people into 'egalitarian communitarians' and 'hierarchical individualists' rather than 'liberals' and 'conservatives'. This is one place where Kahan and collaborators use more nuanced language than I do here, but for my purposes we do not really need the nuance, and I decided it was best to use the same terms here as I have elsewhere in this book.

- 37 per cent of conservatives think most scientists agree that there are safe methods of geologically isolating nuclear waste (burying it underground), while only 20 per cent of liberals think most scientists agree. (35 per cent of liberals think most scientists disagree and 45 per cent on both sides of the political spectrum think scientists are divided.)
- 47 per cent of conservatives think most scientists agree that permitting adults without criminal records or histories of mental illness to carry concealed handguns in public decreases violent crime, whereas only 10 per cent of liberals think most scientists agree. (47 per cent of liberals think most scientists disagree and similar percentages on both sides of the political spectrum think the scientists are divided.)

But what do these findings tell us? It is often claimed that there are important cognitive *asymmetries* between liberals and conservatives. Put bluntly, liberals and conservatives form views in different ways—one side uses reason and evidence, the other . . . not so much (Hodson and Busseri 2012; Kanazawa 2010). It may well be that there are some asymmetries. But these findings suggest that there are also some important *symmetries* (Gampa et al. 2019). Both liberals and conservatives form beliefs about scientific issues in ways that accord with their political convictions. In doing so, they end up at odds with expert scientific opinion, at least as it is represented in expert consensus reports from the US National Academy of Sciences (NAS). The NAS has issued reports with the following conclusions:

- There is scientific consensus that global warming is real, and that human activity is a central cause of it (National Research Council Committee on Analysis of Global Change Assessments 2007).
- There is scientific consensus that there are safe methods of burying nuclear waste (National Research Council Board on Radioactive Waste Management 1990).
- The available evidence does not permit forming a conclusion on the efficacy of concealed carry laws (National Research Council Committee to Improve Research Information and Data on Firearms 2004)

This suggests that conservatives are out of step with the scientific consensus on global warming, while liberals are not. But liberals are out of step with the scientific consensus on the safety of nuclear waste disposal. And both sides overestimate the extent to which the evidence supports their preferred position on concealed carry laws. Now, it may be claimed that these findings still support the conclusion that conservatives are *more* out of step with the scientific consensus than liberals. Perhaps politically motivated reasoning is more pronounced (or at least more damaging) amongst conservatives than liberals. This might be true, but it hardly shows that liberals don't also engage in politically motivated reasoning, or that, in doing so, they don't regularly end up at odds with expert scientific opinion. (For relevant discussion, see Baron and Jost 2019; Ditto et al. 2019.)

This completes my recap of the literature on politically motivated reasoning. I will now turn to its sceptical import. In the following sections, I argue that beliefs that are formed through politically motivated reasoning are epistemically suspect. I start by arguing that they are epistemically suspect because politically motivated reasoning is not a reliable way of forming beliefs about these sorts of topics.

8.2 The Unreliability of Politically Motivated Reasoning

Here is the argument:

(1) Many of us form beliefs about scientific topics that are politically relevant (like global warming) through politically motivated reasoning.
(2) Politically motivated reasoning is not a reliable way of forming beliefs about these sorts of topics.
(3) If the way in which you have formed a belief (or set of beliefs) is not reliable, then the belief(s) in question are not justified.
(4) Many of us do not have justified beliefs about scientific topics that are politically relevant.

I will defend each premise in turn. First, the empirical evidence recapped in §8.1 shows that many of us form beliefs about politically relevant scientific topics through politically motivated reasoning. That said, it is worth emphasizing some respects in which the influence of politically motivated reasoning is limited.

As I have mentioned in previous chapters, financial incentives, even very small ones, can diminish the impact of politically motivated reasoning on judgements about political issues (Bullock, Gerber, and Hill 2015; Prior, Sood, and Khanna 2015). Further, politically motivated reasoning is most pronounced when it comes to attitudes and beliefs to which we attach a lot of importance, whether personal, political, or cultural (Howe and Krosnick 2017). Finally, experts don't engage in politically motivated reasoning with respect to issues within their domain(s) of expertise (Kahan et al. 2016). This last point is important: my argument does not have the implication that, for example, climate scientists do not have justified beliefs about global warming and its likely impacts.

The crucial point for my purposes is that there is no evidence that the effects of politically motivated reasoning lessen as scientific comprehension and literacy increase. On the face of it, you would think that public scepticism about what are regarded as settled issues in science (like that human activity is causing global warming) is correlated with public ignorance about the relevant science. Put bluntly, the less you know about climate science, the more likely you are to be sceptical about the existence of global warming, still less that human activity is causing it. But Taber and Lodge's (2006) finding that subjects who are knowledgeable about a political issue are no less likely to engage in politically motivated reasoning holds for subjects who are knowledgeable about a scientific issue too. One study found that, in general, knowing more about a scientific issue doesn't make you any less likely to engage in politically motivated reasoning and, for conservatives, higher levels of scientific comprehension are actually associated with a small decrease in the perceived seriousness of the threat posed by global warming (Kahan, Peters et al. 2012).

Second, the empirical evidence also suggests that politically motivated reasoning is not a reliable way of forming beliefs about politically relevant scientific topics. Politically motivated reasoning leads liberals to form true beliefs about some scientific topics (e.g. global warming). But it also leads them to form false beliefs about other scientific topics (e.g. the safety of burying nuclear waste). Further, politically motivated reasoning leads conservatives to form false beliefs about global warming, though it also means they are less out-of-step with the scientific consensus on the safety of burying nuclear waste. These findings suggest that politically motivated reasoning is a way of forming beliefs about politically relevant scientific topics that sometimes gets it right, but often gets it wrong. But that is just to say that it is not a reliable way of forming beliefs about such topics.

You might object that I am relying on quite a small set of topics in arriving at this conclusion—too small to permit the formation of a definitive conclusion about the reliability of politically motivated reasoning in the relevant domains. But let me say two things here. Firstly, I am *not* claiming that politically motivated reasoning is, in general, an unreliable belief-forming process. Whether it is a reliable way of forming beliefs about a particular topic or in a certain domain is an empirical question and can only be answered by looking at the available empirical evidence. My claim is just that the evidence suggests that politically motivated reasoning is not a reliable way of forming beliefs about politically relevant scientific topics. (You may be able to extend this to other domains. It would be interesting to consider the implications of motivated reasoning for certain forms of self-knowledge.)

Secondly, one of Kahan's key findings is that beliefs about a disparate range of topics (global warming, nuclear power, firearms, nanotechnology, genetically modified organisms, vaccinations) 'pattern together' in partisan ways. Knowing someone's political and broader cultural identity provides a good basis on which to predict their views about these and other issues (see also Kahan et al. 2015, 2017; Kahan, Jenkins-Smith, and Braman 2011; Kahan et al. 2009, 2010). Now, one explanation for this would be that one side of these debates has got all the issues right while the other has got all the issues wrong—they have demonstrated a reliable capacity to form true beliefs about a wide range of disparate and complex issues at the intersections of science and politics. Another explanation would be that motivated reasoning has led both sides to ignore relevant evidence when it suits them politically. While both explanations are too crude to be the whole story, the claim I am relying on here is just that the latter explanation is closer to the truth than the first (for a detailed defence of this claim, see Joshi 2020).[2]

Where the first two premises require empirical support, the third premise requires epistemological support. The second premise fits nicely with a reliabilist picture of justified belief, on which, put roughly, beliefs are justified if and only if they are produced by reliable belief-forming processes (Goldman 1979). On this picture, if a belief is not produced by a reliable belief-forming process, it is unjustified.

Note, however, that the third premise does *not* say that a belief that is produced by a reliable belief-forming process is thereby justified. Strictly speaking, the third premise is only committed to the necessity of reliability for justified belief, not the sufficiency. This is important because a standard objection to reliabilism as a view of justified belief is that it seems possible for a subject to have a belief that is produced by a reliable process yet lack any awareness not only of this fact, but more generally of how they came by the belief. Many think that, in such cases, the subject's belief is intuitively not justified even though it satisfies the reliabilist's condition for justified belief (see Bonjour's 1980 clairvoyance case). While I am not entirely convinced by this objection (see Chapter 6), for my purposes here I can grant the point.

So far, I have been arguing that, taken individually, each premise of the argument is plausible. Addressing some objections should help to clarify what I have said so far, and to identify what I take to be the most promising lines of resistance to the argument.

The first objection is that a belief may be initially formed via a process that is not reliable, yet the believer later comes to have independent evidence for thinking the belief is true. Imagine that Catriona formed the belief that Morven is innocent of a terrible crime because of wishful thinking, but she later gets excellent evidence that Morven is innocent and believes that she is innocent on this basis. You might think that, while Catriona was not initially justified in believing that Morven is innocent, her belief becomes justified once she begins to hold it based on this evidence. Similarly, you might accept that beliefs formed via politically motivated reasoning are not reliable and so not justified at the point of acquisition but also

---

[2] One way in which things are more complicated than this is that, at least with respect to many of these issues, there are not just two sides. For example, what is the 'left-wing' stance on nuclear power? What is the 'right-wing' stance on vaccinations? The answer is that it depends on many other variables (the country, the vaccine, the type of right- or left-wing ideology). Still, it may be that knowing which of these many groups someone belongs to helps predict their stance on a wide range of issues. If this is right, then this strengthens the argument in the main body of the text. It would have to be that just one of these many groups has managed to form true beliefs about most of these issues. What are the chances of that?

hold that a belief that was formed via politically motivated reasoning can become justified because the believer (the person with the belief) may later acquire excellent evidence for the belief and come to hold it on the basis of this evidence.

I am happy to acknowledge that, if someone who has formed a belief through politically motivated reasoning later comes to have good independent evidence that the belief is true and comes to hold the belief based on this evidence, then their belief is justified. The crucial point is that this is not what the politically motivated reasoning paradigm says normally happens. Politically motivated reasoning impacts not just on how we initially form beliefs but on how we assess new evidence and information. When we rely on expert opinion—perhaps as a means of settling issues about which we recognize we are ignorant—our assessments of who the experts are is also driven by politically motivated reasoning.

If this is right, then it is not just that a global warming sceptic (or a sceptic about the safety of nuclear power) initially forms a view about global warming that conforms with their political beliefs and broader cultural background. They assess new evidence and information, including expert testimony, in ways that buttress their existing (sceptical) beliefs. As a reliabilist would put it, it is not just that the process by which they initially formed the belief was not reliable. The process by which they *maintain* the belief is not reliable either. So we can run this further argument:

(1) Many of us maintain our beliefs about scientific topics that are politically relevant (like global warming) through politically motivated reasoning.
(2) Politically motivated reasoning is not a reliable way of maintaining beliefs about these sorts of topics.
(3) If the way in which you maintain a belief (or set of beliefs) is not reliable, then the belief(s) in question are not justified.
(4) Many of us do not have justified beliefs about scientific topics that are politically relevant.

This argument is supported by similar considerations to the previous argument.

I have been talking about politically motivated reasoning as a belief-forming process. The second objection is that this is not necessarily the right description of the processes by which individuals go about forming beliefs about political issues, and issues at the intersection of science and politics like global warming. This brings to the fore one of the standing problems with reliabilism, which is the so-called 'generality problem'. Briefly: the reliabilist holds that a token belief is justified if and only if it is produced by a reliable belief-forming process. But *types* of belief-forming process are properly assessable in terms of reliability, not tokens, and any token belief-forming process will be a token of several different belief-forming process types which may differ when it comes to their reliability (Conee and Feldman 1998).

As it arises here, the problem is that the token beliefs I am claiming are the result of politically motivated reasoning can be seen as the products of a range of different belief-forming process types that may differ in terms of reliability. For instance:

- Beliefs formed through politically motivated reasoning.
- Beliefs formed through 'liberal politically motivated reasoning'.
- Beliefs formed through 'conservative politically motivated reasoning'.

This list could be continued indefinitely but these three possibilities suffice to illustrate the problem: it may be that beliefs formed through liberal politically motivated reasoning are reliable, whereas beliefs formed through conservative politically motivated reasoning are not, or vice versa.

I want to make two points in response. The first is to reiterate what I said earlier: the evidence marshalled in §8.1 makes it plausible that both conservative *and* liberal politically motivated reasoning are not reliable belief-forming processes. You could justifiably claim that conservative politically motivated reasoning is *less* reliable than liberal politically motivated reasoning. But this hardly shows that liberal politically motivated reasoning is reliable. At the very least, the onus is on my opponent to make the case that this is so.

Of course, this does not rule out finding some other way of carving out 'good' cases of politically motivated reasoning (cases where it leads to the formation of true beliefs) and distinguishing them from 'bad' cases (cases where it leads to the formation of false beliefs). This is where my second point comes in. I do not claim—or need—to have a solution to the generality problem. But it is reasonable to adopt the following rough-and-ready heuristic. When psychologists, or empirical scientists in general, identify a class of beliefs as products of a particular belief-forming process type, this is (defeasible) reason to treat the relevant process type as the one that is relevant for the purposes of determining reliability.

I take this heuristic to fit with the broader naturalistic aspirations of reliabilist epistemologies, as outlined in Goldman's 1986 classic *Epistemology and Cognition*. In the empirical work on politically motivated reasoning, a general process-type— politically motivated reasoning—is identified as the (or at least a central) cause of beliefs about certain contentious scientific topics. Further, no distinction is made between the process by which liberals individuals form such beliefs and the process by which conservatives form such beliefs. I take this to be (defeasible) reason to focus on politically motivated reasoning in general, and not to distinguish between how it manifests itself in individuals with different political convictions.

Finally, what if you do not buy the claim that a belief needs to be produced (or maintained) by a reliable belief-forming process to be justified? What if you are attracted to the view that whether a belief is justified depends on whether it is supported by the *evidence* that you have at your disposal? On this picture of justified belief—a picture known as *evidentialism*—what matters is your evidence and whether your beliefs are supported by it, not reliability (Conee and Feldman 2004).

Of course, we tend to think that believing based on the available evidence is a reliable way of forming beliefs. But, in cases where the available evidence is itself misleading, believing based on it will not be a reliable way of forming beliefs. Where the reliabilist claims that, in such cases, the beliefs in question are not justified, the evidentialist may claim that they are justified because they are supported by the available evidence, even though that evidence is misleading. (What exactly they say depends on their views about evidence gathering and whether a failure to gather enough or the right sort of evidence impugns the justificatory status of beliefs based on said evidence. See Begby 2021, ch. 1 for discussion.)

This objection is important because you might think that, when it comes to beliefs formed through politically motivated reasoning, it often happens that we have misleading evidence. As I have emphasized, when we engage in politically motivated reasoning, we gather evidence to support views on politically relevant issues that accord with our social and political values. As a result, it is plausible that, on an evidentialist picture of justified belief, beliefs formed through politically motivated reasoning will (often) be justified (Levy 2021). I myself prefer the reliabilist picture to the evidentialist picture (recall Chapter 6). But I appreciate that the reader may be more sympathetic to evidentialism than I am. More generally, I do not want my argument in this chapter to stand or fall with the viability of (something like) the reliabilist picture of justified belief. In the next section, I offer an argument for the same conclusion as the argument in this section, but which should be more acceptable to those with evidentialist sympathies.

8.3 Politically Motivated Reasoning and Basing

Consider the following pair of cases:

NAÏVE STUDENT-1:   Tim is an impressionable student and is inclined to blindly trust socially approved opinions with little to no critical scrutiny. Fortunately for Tim, he is raised in a household where exposure to and social praise of sound climate science is abundant, and 'global warming denial' is discussed only in a negative light. Tim, like most around him, believes that global temperatures are rising, and that human activity is the cause.

NAÏVE STUDENT-2:   Tim* is an impressionable student and is inclined to blindly trust socially approved opinions with little to no critical scrutiny. Unfortunately for Tim*, he is raised in a household where exposure to and social praise of sound climate science is rare, and 'global warming' is discussed only in a negative light. Tim*, like most around him, believes that it is *not* the case that global temperatures are rising.

With the literature on politically motivated reasoning in hand, we can explain what is going on in these cases. Both Tim and Tim* assess evidence about climate science in light of their background political beliefs and broader cultural backgrounds. But, because these background beliefs are quite different, these assessments lead them in opposite directions. Tim ends up believing that global temperatures are rising, and that human activity is the cause. Tim* ends up denying this. I submit that neither Tim nor Tim* is particularly unusual. While my descriptions of both Tim and Tim* are undeniably simplistic, many (though of course not all) individuals are, in most essential respects, like Tim or Tim*.

What are we to say about the justificatory statuses of Tim's and Tim*'s beliefs? A resolute evidentialist might argue that they are on a par, at least with respect to justification. Both Tim and Tim* form their beliefs based on the available evidence so both Tim's and Tim*'s beliefs are justified. Of course, we would want to say that Tim*'s evidence is misleading, and Tim*'s beliefs are false, whereas Tim's evidence is not misleading, and his beliefs are true. But, for the resolute evidentialist, this need not make a difference to the justificatory status of their beliefs.

A less resolute evidentialist might try to draw a distinction between Tim and Tim*. Tim has excellent evidence for his beliefs because he happened to grow up in an epistemically friendly environment. Tim* has not got good evidence for his beliefs because he happened to grow up in an unfriendly environment, at least as far as climate science is concerned. As a result, Tim's beliefs are justified while Tim*'s are not (alternatively, Tim's beliefs are more justified than Tim*'s, or are justified in a way in which Tim*'s are not). Of course, it would be unfair to blame Tim* for this—he is not responsible for the environment he grew up in. Still, this does not make his beliefs justified.

I want to argue for a different take on the situation: Tim's and Tim*'s beliefs are on a par, but that is because neither of them has justified beliefs. This is because, put briefly, even if they have good evidence for their beliefs, neither of them bases his belief on this good evidence in the way required for those beliefs to be justified.

My argument for this claim starts with the standard epistemological distinction between *propositional* and *doxastic* justification (Alston 1985; Korcz 2000; Kvanvig 2003; Pollock and Cruz 1999). This is the distinction between having good evidence for your belief (propositional justification) and properly basing your belief on the good evidence you have (doxastic justification). The idea then is as follows. Let us grant that Tim has good evidence for his beliefs. Whether Tim* has good evidence for his beliefs depends on your theoretical commitments (what is good evidence?). For the sake of argument let us grant that his evidence is good too. Even granting this, it is not clear that either Tim or Tim* bases his belief on the good evidence they have (or take themselves to have). Rather, it seems that, to the extent that they base their beliefs on the good evidence they have, they only do this because their beliefs and this evidence has the epistemically irrelevant property of being socially approved. If that is right then, while their beliefs may be propositionally justified, they are not doxastically justified because they do not base their beliefs on the good evidence they have.

If this argument works, then beliefs formed through politically motivated reasoning are unjustified simply in virtue of the fact that they are not based on the evidence that the believer has. Importantly, this argument takes no stand on what makes for good evidence so it should be acceptable no matter your other epistemological commitments. But does it work? Let me develop the argument in more detail by considering two objections. In what follows I will just talk about Tim for reasons of simplicity. (It would be odd indeed to hold that Tim*'s beliefs are justified while Tim's are not.)

First, I have argued that Tim's beliefs are (doxastically) unjustified because, to the extent that he bases these beliefs on the good evidence he has, he does this because this evidence has the epistemically irrelevant property of being socially approved. You might object that this property is not epistemically irrelevant. Tim is using an effective strategy for deciding what to believe, which is to defer to socially recognized authorities. Deferring to socially recognized authorities is not a bad epistemic strategy. It is not going to work in all environments—socially recognized authorities can be wrong (see Tim*). But good inferential strategies need not work in all possible environments (Gigerenzer 2000).

This objection can be dealt with briefly. I agree that there need not be anything wrong with deferring to recognized authorities, and that good inferential strategies need not work in all environments. But I also do not think this causes any trouble for my argument.

Let us start with the first concession (that there need not be anything wrong with deferring to recognized authorities). The crucial question is what this deference is based on. Deferring to recognized authorities because you have evidence that they are likely to have the right answers (or are more likely to have them than you are or anyone else is) is a good epistemic strategy. But deferring to recognized authorities because you recognize that they share certain core values with you, or belong to the same social group as you, looks less like a good epistemic strategy. The mere fact that you share certain core values with someone surely cannot be evidence that they are likely to have the right answers, at least on factual matters like global warming. But this is exactly what the politically motivated reasoning paradigm says we do: we defer to (what we recognize as) authorities precisely because we recognize that they share our values.

What about the second concession (that good inferential strategies need not work in all environments)? The problem with the strategy of deferring to those with whom you share certain core values is not that there are environments in which it is unreliable. The problem is that, at least for many of us, it is unreliable in the world in which we find ourselves. This goes back to the point I made in the previous section: both liberals and conservatives have views that depart from the scientific consensus. As before: I do not claim to have made a conclusive case for this claim. But I do think that the onus is on my opponent to show that I am wrong about this.

Second, you might object that Tim does properly base his belief on the evidence he has. Compare Tim with an individual who does not properly base his belief on the evidence he has, Jennifer Lackey's 'racist juror':

RACIST JUROR: Martin is a racist juror. He receives, over the course of the trial, compelling testimony that the defendant is guilty. Martin, however, bases his belief that the defendant is guilty not on the good reasons he has for thinking so, but on his racist belief that individuals of the defendant's ethnicity are likely to commit criminal acts. (This is a variant on a case in Lackey 2007)

Both Tim and Martin have beliefs that are propositionally justified: there is good evidence supporting the relevant propositions. But, as far as doxastic justification is concerned, there seems to be a crucial difference between Tim and Martin. While Tim bases his belief that scientists agree that global temperatures are rising on the good evidence that he has, Martin bases his belief that the defendant is guilty on bad evidence (racist, presumably unjustified, beliefs). Granted, Tim bases his belief on this good evidence *because* it is socially approved in his environment. But he still bases his belief on this evidence. So, we should maintain that, unlike Martin, Tim is not only propositionally justified but also doxastically justified.

Dealing with this objection will require saying more about the basing relation and the relationship between basing and doxastic justification. What follows is a little involved and the reader who does not see the force of the objection in the first place might want to skip to §8.4.

The second objection is based on the claim that basing a belief on good evidence *suffices* for doxastic justification in the sense that, if a subject S bases her belief that p on evidence E, and E is good evidence for believing p, then S is doxastically justified in believing that p. If basing a belief on good evidence suffices for doxastic justification in this way, then it follows that Tim's belief is doxastically justified. (Martin's belief is not doxastically justified because it is not based on the good evidence that in fact supports it.) But I am going to argue that basing a belief on good evidence is not sufficient for doxastic justification, even if it is necessary. To do this I will distinguish between three 'families' of accounts of the basing relation: doxastic accounts, inferential accounts, and causal accounts. I will take each family in turn and argue that defenders of each account of basing need to hold that merely basing a belief on good evidence is not sufficient for the belief to be doxastically justified.

First, on a doxastic account of the basing relation, basing a belief on evidence is a matter of possessing a meta-belief to the effect that the evidence is a good reason for holding the belief (Audi 1982; Ginet 1985). The idea is that your belief is based on a piece of evidence when you have the additional belief that the evidence in question is a good reason for holding the belief in question. Your belief is based on good evidence when you have the meta-belief that the evidence is a good reason for holding the belief, and the evidence is in fact a good reason for holding the belief.

The problem though is that, if the meta-belief is itself the product of something like wishful thinking, then the first-order belief is surely not doxastically justified. Imagine I believe that I am well-liked by my colleagues and have good evidence that this is so. I also believe, of my first-order belief that I am well-liked, that it is based on good evidence. Unfortunately, this meta-belief is itself a product of my desire to have justified beliefs. (Maybe I want to have justified beliefs in general, or I just want to have justified beliefs about how well-liked I am.) In this case, my first-order belief is surely not doxastically justified, even though I have good evidence that I am well-liked. If my first-order belief were justified, then my desire to have justified beliefs would be part of the reason I have justified beliefs. But this is implausible. So defenders of doxastic accounts of basing need to hold that basing a belief on good evidence is not sufficient for doxastic justification. It is not sufficient because you might base your belief on good evidence for bad reasons.

Second, on inferential accounts of the basing relation, basing is to be understood in terms of inference (Bondy and Carter 2019). The idea is that your belief is based on a piece of evidence when you infer (if only implicitly) the belief from the evidence. So, your belief is based on good evidence when you infer the belief from the evidence and the evidence is in fact good.

The problem here is that there is a crucial difference between competent and incompetent inference. How exactly you cash out this difference will depend on your other theoretical commitments. But there is an intuitive difference between 'good' inferences (e.g. logically valid inferences) and 'bad' inferences (e.g. fallacious inferences). Imagine I infer that the light is broken from (i) the premise that, if the light is broken, then the room will be dark, and (ii) the premise that the room is dark. This inference is fallacious (I have affirmed the consequent). According to an inferential account of basing, my belief (that the light is broken) is based on these premises. But it is surely not thereby doxastically justified, even if the premises are in fact good evidence for the conclusion (perhaps the best explanation of why the room is dark is in fact that the light is broken). So defenders of inferential accounts of basing also need to hold that basing a belief on good evidence is not sufficient for doxastic justification. It is not sufficient because you might infer the belief from the good evidence but via a fallacious or otherwise bad inference.

Third, and finally, on causal accounts basing is a matter of causation in the sense that a subject S's belief that p is based on evidence E iff S's belief that p is caused or causally sustained by E (Harman 1970; McCain 2012; Moser 1989; Swain 1981; Turri 2011). The idea is that your belief is based on a piece of evidence when it is the evidence that caused or sustains the belief. Your belief is based on good evidence when it is caused or sustained by good evidence. (We need to add something to ensure that the causal chains are not deviant, but we can set this aside here.)

The problem again is that, even if a belief is caused or sustained by good evidence, it may not be doxastically justified. Imagine that, whenever I see that the streets are wet, I form the belief that it has been raining: my evidence that the streets are wet causes (and/or sustains) my belief that it has been raining. But I do this because of a superstition, not because I recognize or understand the connection between wet streets and rain. This superstition leads me to form lots of other, false beliefs. For example, when I see that the streets are wet, I also form the belief that I will have an unpleasant day, which is usually not true (my days are quite pleasant). In other words, the way in which my belief is caused is unreliable, though it usually yields true beliefs about whether it has rained. So, defenders of causal accounts need to hold that basing a belief on good evidence is not sufficient for good evidence too.[3]

The take-home message is that doxastic justification requires not just basing a belief on good evidence but *good* or *proper* basing, where being based on good evidence is necessary but not sufficient for proper basing. Of course, what else proper basing requires is another question. But here is not the place to address it. All I need for my purposes is the point that doxastic justification requires proper basing. If this is right, then the fact that Tim bases his belief on good evidence in a way that Martin (the racist juror) does not is insufficient to show that Tim properly bases his belief on good evidence, and so it is not enough to show that his belief is doxastically justified.

You might wonder if I can also offer a positive argument that Tim's belief is not doxastically justified. I can. Tim's case is analogous to a case where it seems clear that the relevant belief is not doxastically justified. Here is the case:

FRANCOPHILE CARTOGRAPHER: Rae irrationally believes that French cartographers are the *only* reliable sources of cartographical information. Rae's Francophilia so strongly influences her assessment of the reliability of maps that she distrusts all information written by non-French authors. Rae stumbles upon several pieces of evidence (E1, E2, and E3) for believing cartographical claim X. But, simply because E1, E2, and E3 were written by Italian cartographer Giacomo

---

[3] I recognize that my discussion of causal accounts of basing raises worries about the generality problem. I suspect that these worries parallel the worries I discussed in §8.2 so I will not say anything more about this issue here.

Gastaldi, Rae disregards this evidence and so does not come to believe claim X because of it. Later that day, Rae encounters the *very same* pieces of evidence (E1, E2, and E3) for cartographical claim X, but this time French cartographer Pierre Desceliers wrote them. Because—and only because—Desceliers is French, she accepts this evidence, and comes to believe X because of it.

Rae bases her belief on what we may assume is good evidence that she has for believing claim X—that is, E1, E2, and E3—but she does not *properly base* her belief on this evidence. That is, she does not base her belief on this evidence in the way that would be required for doxastic justification. When Rae encounters these pieces of evidence in the work of Gastaldi, she disregards them; when she encounters them in the work of Desceliers, she accepts them. Rae's belief is not properly based on these pieces of evidence if whether she believes on the basis of them depends primarily on the nationality of the putative expert, as this is clearly an epistemically irrelevant feature. (That it is epistemically irrelevant is part of the case—we are imagining a case where French cartographers are no more reliable than non-French cartographers.) As a result, her belief is not doxastically justified, even though it is based on good evidence.

Let me spell this out in a bit more detail, starting with doxastic accounts of basing. We can grant that Rae has a meta-belief to the effect that E1, E2, and E3 are good pieces of evidence for X and so she bases her belief that X on this good evidence. But this meta-belief is itself unjustified because it is a result of the irrational prejudice that French cartographers are superior to Italian. Because irrational prejudice surely cannot convert an otherwise doxastically unjustified belief into a doxastically justified belief, it follows that Rae's belief, though based on a good reason, falls short of doxastic justification on a doxastic account of basing.

Turning to inferential and causal accounts, the process by which Rae comes to (and holds on to) her belief is both incompetent and unreliable, even though it issues in a good reason as a basis. The inference from someone being French to their being a reliable cartographer is a bad one, given that it is part of the case that it is based on an irrational prejudice. (The same goes for the inference from someone being non-French to them being an unreliable cartographer.) Similarly, the disposition Rae manifests in forming these beliefs is generally unreliable even if all the French cartographers Rae is in contact with happen to speak truly, and most of the Italian cartographers she is in contact with happen to speak falsely (apart from Gastaldi, of course).[4] Because incompetent and

---

[4] This raises a well-known issue in the literature on reliabilism. Processes that we would want to describe as unreliable will turn out to be reliable in certain environments. Guessing is a paradigm example of an unreliable process but guessing will be reliable if the environment is structured so that guesses are usually correct. For this reason, the kind of reliability that is epistemically important is general, or normal, reliability—reliability in normal environments. For different ways of unpacking the relevant notion of normality, see Graham (2012) and Sosa (2015).

unreliable basing is justification-undermining on causal and inferential accounts, Rae's belief, though based on a good reason, is not doxastically justified on either of these accounts of basing.

The point is that, if we accept this line of thinking, then by parity of reasoning we should also accept that Tim is not doxastically justified in believing that global temperatures are rising. The cases are structurally identical. In both cases, the subject (Rae or Tim) bases their belief on good evidence, but not because it is good evidence, but rather because of something epistemically irrelevant. As a result, in both cases the process by which the subject bases their belief on good evidence does not produce doxastic justification. Importantly—and this is what I have been at pains to stress—this point stands regardless of which view of basing you subscribe to.

The discussion in this section has got a little involved so let me finish by summarizing the main line of argument. There are two ways in which your belief can fail to be doxastically justified even though it may be propositionally justified. You may be like Martin the racist juror and believe some proposition for which there is good evidence, but not because of this good evidence. Or you may be like Tim or Rae and believe some proposition for which there is good evidence, and based on this good evidence, but not *because* it is good evidence, but for some other epistemically irrelevant (or plain bad) reason. As I have argued, while these ways may differ, in both cases the result is the same: the belief in question is unjustified.

Further, while I have focused on an imagined epistemic agent (Tim), he is typical of the sort of epistemic agent who engages in politically motivated reasoning. What I have said about Tim's belief that global temperatures are rising should go for anyone who has formed their belief about an issue like global warming through politically motivated reasoning. These beliefs are not going to be doxastically justified because, even when they are true and held based on good evidence, they are not held based on good evidence because it is good evidence. If this is right, then our sceptical worry remains: beliefs formed through politically motivated reasoning can be epistemically suspect *even when they are true and held based on good evidence*.

This completes my defence of the second argument. Where the first argument assumed that reliability (of a belief forming process) is necessary for justification (of beliefs produced via that process), the second argument appealed to the basing relation. I have argued that, on any account of the basing relation, beliefs formed via politically motivated reasoning are not based on the evidence the believer has at their disposal in the way required for doxastic justification. Importantly, this is so even if the evidence is good evidence. The result is an argument for my main conclusion that should be acceptable even to evidentialists and others who reject a reliabilist picture of justified belief.

I am going to finish this chapter by relating all of this to non-ideal epistemology and my argument in this book. First, though, I want to explain how my argument so far relates to the literature on debunking arguments and irrelevant influences on belief.

8.4 Debunking Arguments

I have been considering the epistemological implications of the fact that many of our beliefs are influenced by 'epistemically irrelevant' factors—factors that have nothing to do with the truth of their contents. The reader who is familiar with the literature on debunking arguments and epistemically irrelevant influences on belief might be wondering how my argument in this chapter relates to this literature (see Avnur and Scott-Kakures 2015; DiPaolo and Simpson 2016; Mogensen 2016; Srinivasan 2015; Vavova 2018; White 2010). The reader who is not familiar with this literature, or is not interested in how my argument relates to it, can skip to the next and final section.

There are two important respects in which my argument in this chapter differs from the bulk of the literature on epistemically irrelevant influences on belief. First, those who think that the fact that many of our beliefs are influenced by irrelevant factors grounds a distinctive sceptical worry about those beliefs tend to frame the worry in terms of *defeat* (e.g. Avnur and Scott-Kakures, DiPaolo and Simpson, Vavova). The idea is that learning, of one or more of your beliefs, that it was influenced by an irrelevant factor might give you a defeater for the belief(s) in question. Katia Vavova, for example, defends this principle:

> Good Independent Reason Principle: To the extent that you have good independent reason to think that you are mistaken with respect to p, you must revise your confidence in p accordingly—insofar as you can.   (2018, p. 145)

Imagine that my political beliefs have been shaped by my upbringing. Further, I come to recognize that this is true. Vavova's thought is that this can generate a sceptical worry about my political beliefs, but it does so only if I have good independent reason to think that my beliefs are likely to be false because they were shaped by my upbringing. It may well be that I have no such reason (perhaps I do not think my upbringing makes my political beliefs any more likely to be false than they were already). If this is the case, then the fact that my political beliefs were influenced by an irrelevant factor is no grounds for scepticism about those beliefs. On the other hand, if I do have a good independent reason (perhaps I recognize that my upbringing makes my political beliefs more likely to be false than they were already) then there are grounds for scepticism.

What Vavova says about the sceptical potential of these sorts of defeaters seems to me to be correct. It is, however, important to recognize that it is not what I have been arguing for in this chapter. I have argued that politically motivated reasoning can ground a sceptical worry about beliefs formed through it for one of two reasons: politically motivated reasoning is not a reliable way of forming beliefs in some domains (the first argument) and beliefs formed through politically motivated reasoning are not properly based on whatever evidence the believer has for them (the second). If these arguments work, they generate a sceptical challenge that applies irrespective of whether those of us who have formed our beliefs through politically motivated reasoning are worried about the epistemic status of our beliefs or have a defeater for them. (A belief can be formed via an unreliable process or not be properly based on whatever evidence you have for it without you having any reasons for thinking this is so.)

This difference is the consequence of a second, more fundamental difference. Vavova, like many who write on irrelevant influences, seems to see the problem posed by irrelevant influences as one that arises from the first-personal perspective of the believer. You are supposed to reflect on the origins of your beliefs and consider whether those origins in any way discredit them. Within this framing, it is natural to focus on the consequences that *recognizing* that your beliefs have been influenced by irrelevant factors has for their justificatory status.

I have, in contrast, approached things from a more external, social psychological perspective. Within this framing, it is more natural to focus on whether social psychological evidence about the likely origins of our beliefs reveals them to be unjustified. As a result, I have not considered whether the subjects I have discussed (e.g. Tim and Tim*) are *aware* of the origins of their beliefs. I have considered psychological evidence about the genesis of their beliefs and argued that they are unjustified because of their genesis. Of course, this does not preclude me from endorsing Vavova's claims about defeat. It may be that *recognizing* that your beliefs were formed via politically motivated reasoning can provide you with a defeater for those beliefs and so ground a further sceptical argument. But this is not the argument I have developed in this chapter.

8.5 Scepticism in Non-Ideal Epistemology

I have argued for a kind of scepticism according to which our beliefs about politically contentious scientific issues like global warming may well be unjustified. I want to finish by saying more about this kind of scepticism. I also want to tie the argument of this chapter into the overall argument of this book. In the process I will address a further objection to my sceptical argument that might seem to arise from a non-ideal perspective.

Let us start with the kind of scepticism that is at issue here. First, my claim is that beliefs that are formed through politically motivated reasoning are not (doxastically) justified. Given the ubiquity of politically motivated reasoning, particularly when it comes to views about politically relevant scientific issues, we can conclude that many of our beliefs about such issues are not justified. Further, most epistemologists hold that, if a belief is not doxastically justified, it does not count as knowledge. If the argument works, then, it establishes the further conclusion that many of us do not have knowledge about these issues. (Some epistemologists deny that doxastic justification is necessary for knowledge, but this is typically because they think you can have knowledge that is not based on evidence in the first place. See Millar 2019.)

Second, this kind of scepticism is local rather than global. It concerns our beliefs about certain issues—scientific issues that are politically relevant. This may not be a class of beliefs that is important from a purely philosophical point of view. What, exactly, do the members of the class have in common? One is tempted to say: nothing, beyond the fact that they are scientific issues that for whatever reason have become politically relevant. But it is clearly a class of beliefs that is important from a less narrowly philosophical point of view. For one thing, it is a class of beliefs that is consequential because our beliefs about these issues inform our political choices and decisions (recall Chapter 4).

Third, this kind of scepticism is based on empirical evidence, not philosophical reflection or speculation. In this respect it differs markedly from more familiar forms of scepticism that are arrived at via philosophical argument (e.g. Cartesian scepticism). Still, it is important to recognize that my arguments in this section have relied on substantive philosophical claims. I have relied on (something like) a reliabilist picture of justified belief (the first argument) and some claims about basing (the second argument).

Fourth, this kind of scepticism is contingent in the sense that the claim is not that we *could not* have justified beliefs/knowledge about these issues but that, in fact, many of us do not. While there may be good reasons why particular individuals are highly likely to engage in politically motivated reasoning and other individuals are less likely, these reasons hardly entail that anyone *must* engage in politically motivated reasoning. More generally, they certainly do not entail that knowledge or justified belief about these issues is impossible. It is clearly possible. Indeed, we can safely assume that a few of us have it (e.g. climate scientists).

Fifth, and relatedly, it is a kind of scepticism which applies to some agents (those who have formed their beliefs about the relevant issues through politically motivated reasoning) but not others (those who have not). While it is standard to consider forms of scepticism that are domain-specific (e.g. scepticism about our empirical beliefs, scepticism about our moral beliefs, scepticism about our beliefs about the future) it is more unusual to consider a form of scepticism that applies to some epistemic agents but not others. While I admit it is unusual, this is a consequence of the fact that sceptical arguments are usually based on philosophical considerations about the nature of knowledge or justification which, by their very nature, apply either to all agents or to nobody rather than empirical considerations which, by their very nature, are less universal.

In view of all this, the kind of scepticism I have been arguing for can sensibly be viewed as a kind of scepticism that belongs within a non-ideal epistemology. It is based on considerations about human psychology, not thought experiments and philosophical reflection. This kind of scepticism is not entirely new. You can find a very clear statement (if not an unambiguous endorsement) of it in some work in naturalized epistemology (see especially Kornblith 1999). It is arguably the sort of scepticism that is at issue in the 'rationality wars' in psychology (for discussion, see Samuels, Stich, and Bishop 2002 and Sturm 2012). But it is fair to say that it lacks an accepted name. Following Carter and Littlejohn (2021), I propose calling this the 'new scepticism'. By calling it 'new' I intend to contrast it with older, more familiar forms of scepticism, such as those based on considerations about evil demons and brains in vats, not to claim it as my own invention.

I now want to consider an objection to my sceptical argument. You might think that the problem with scepticism, whether new or old, is that it sets the 'bar' for knowledge (or justified belief) too high. For example, you might hold that the sceptic about knowledge of the external world sets the bar for knowledge about the external world too high. They may succeed in showing that we cannot meet this high bar, but the lesson is that it was a mistake to have thought we needed to meet this bar in the first place. Similarly, you might hold that, in the case of knowledge (or justified belief) about the sorts of scientific and political issues I have discussed, the 'new sceptic' is setting the bar too high. Our beliefs do not need to be properly based on the available evidence to be doxastically justified. It is enough that they are based on the available evidence, whether properly or not.

As stated, this objection is a bit crude, both in the case of knowledge about the external world and in the case of knowledge about science and politics. But a far more sophisticated version of it has been developed by Robert Pasnau in his 2017 book *After Certainty*. I will briefly explain Pasnau's position, before commenting on what it might mean for my sceptical argument and for the new scepticism.

Let me start with two comments about Pasnau's aims in his book. First, *After Certainty* is an intriguing blend of history of philosophy and philosophical methodology. Pasnau's main aim is to explain how fallibilism came to occupy a dominant position in epistemology, and this is different from endorsing the reasons and methodological shifts that he thinks explains the almost wholesale adoption of fallibilism. Still, Pasnau does seem to endorse some of these reasons, and the methodology that he thinks led to the widespread adoption of fallibilism.

Second, Pasnau calls this methodology 'idealized epistemology'. In the context of this book, this label is liable to mislead. Pasnau is not engaging in ideal epistemology in the sense in which I have used the label. He is engaging in what in Chapter 2 I called the 'theory of epistemic ideals'. His fundamental question concerns the attainability of knowledge viewed as an epistemic ideal:

Rather than take as its goal the analysis of our concept of knowledge, an idealized epistemology aspires, first, to describe the epistemic ideal that human beings might hope to achieve and then, second, to chart the various ways in which we commonly fall off from that ideal. As one might expect, it turns out to be fairly easy to characterize what we would ideally like to achieve in principle and quite hard to come to grips with what we might actually be able to achieve in practice.

(2017, p. 3)

In my terms, Pasnau purses the theory of epistemic ideals, but with a strong emphasis on feasibility constraints. His approach is therefore a sort of blend of ideal and non-ideal epistemology.

Pasnau's basic approach to the theory of epistemic ideals is *iterative*. We start by identifying an epistemic ideal. We then ask whether that ideal is attainable. If it turns out that it is unattainable, we need to find a new, more modest, goal to put in its place. The historical part of Pasnau's story explains the emergence of fallibilism in these terms. Traditionally, the goal of inquiry was held to be certainty; consider, for instance, the role of certainty in Cartesian epistemology. But, starting with Locke, there was growing recognition that, at least as far as beliefs about empirical matters are concerned, certainty is unattainable.

On Pasnau's telling, it was this recognition that led to the emergence of fallibilist conceptions of knowledge—conceptions on which knowledge can be had even though the grounds for empirical beliefs do not supply certainty. For the fallibilist, even though we cannot aspire to certainty, we can still distinguish between beliefs that are held on grounds strong enough for knowledge and beliefs that are not.

We might try to apply Pasnau's approach to the sceptical argument I have developed in this chapter. The idea would be that what my argument really shows is that we need to rethink an epistemic ideal which, though not as central as an ideal like certainty, is still integral to a common way of thinking about knowledge. What is this ideal? It is an ideal concerning the origins of our beliefs. It says, roughly, that we should aspire to have beliefs whose origins are not (potentially) a source of discredit. The suggestion, then, is that the sceptical arguments I have developed in this chapter show the unattainability of this ideal, at least for most of us, and with respect to our beliefs about scientific issues with a political relevance.

You might question whether extending Pasnau's approach in this way is viable. Perhaps his approach makes sense in the case of central epistemic ideals like certainty but makes little sense when it comes to less central, less important ideals, like an ideal concerning the origins of our beliefs. But my objection to Pasnau's approach—considered as a claim about epistemological methodology, not an explanation of a historical fact about the development of our epistemic ideals— is more fundamental. Why would the mere fact that a goal is unattainable be a reason to reject it?

As I have repeatedly emphasized, it seems clear that the mere fact that a goal is unattainable is no reason not to strive for it. A goal may be unattainable yet still functional as the goal of a practice designed around it. Our moral and political ideals may well be like this. Justice—or at least the Platonic ideal of it—may not be attainable. But that is not in itself a reason not to strive for it. The mere fact that, if the sceptic is right, knowledge (whether in general or within a particular domain) is unattainable is not by itself a reason not to aim for it or to replace it with some more attainable goal. Of course, this is not to say that the sceptic is right or that we should aim for the goal in question. It may be that the sceptical argument fails for another reason or that there are other reasons not to aim for this goal.

If this is right, you cannot reject my sceptical argument on the grounds that it presupposes an ideal about the origins of our beliefs that we will frequently fall short of. What would need to be shown is that, in trying to ensure our beliefs are properly based on the available evidence, we run the risk of worsening our epistemic situation. But do we have any reason to think that someone who tries to form properly based beliefs will do a worse job than someone who does not? I do not think that we do, and nothing I have said in this book suggests otherwise.

In closing, let me just draw together a few threads running through this chapter, and the book. In this chapter, I have argued that the empirical literature on motivated reasoning provides the basis for an argument for a distinctive kind of scepticism that I have called the 'new scepticism'. The new scepticism differs from the old in that it is driven by empirical work on the causes and drivers of our beliefs. Moreover, I have used the new scepticism to buttress two points I have been keen to emphasize throughout the book. Contrary to what some may assume, non-ideal epistemology is not necessarily 'less demanding' than ideal epistemology. Relatedly, the non-ideal epistemologist's fundamental point is not that the ideal epistemologist traffics in norms of inquiry and ideals that are hard to follow or hard to attain. If non-ideal epistemology ends up being less demanding than ideal epistemology in some respects, this is not because there is necessarily anything wrong with being demanding. It is because, as it turns out, demanding goals and norms of inquiry tend to worsen rather than improve our epistemic situation. When the ideal epistemologist proposes goals and norms that do this, the only remedy is non-ideal epistemology.

References

Ahlstrom-Vij, K. (2013) *Epistemic Paternalism: A Defence*. Basingstoke: Palgrave Macmillan.

Alfano, M. (2012) 'Expanding the Situationist Challenge to Responsibilist Virtue Epistemology', *Philosophical Quarterly*, 62(247), pp. 223–49.

Alfano, M. (2014) 'Extending the Situationist Challenge to Reliabilism about Inference', in A. Fairweather and O. Flanagan (eds) *Virtue Epistemology Naturalized: Bridges between Virtue Epistemology and Philosophy of Science*. Dordrecht: Synthese Library, pp. 103–22.

Alfano, M. and Fairweather, A. (2017) *Epistemic Situationism*. Oxford: Oxford University Press.

Alston, W. (1985) 'Concepts of Epistemic Justification', *The Monist*, 68(1), pp. 57–89.

American Academy of Arts & Sciences (2018) *Perceptions of Science in America*. Available at: https://www.amacad.org/publication/perceptions-science-america.

Anderson, E. (1995) 'Knowledge, Human Interests, and Objectivity in Feminist Epistemology', *Philosophical Topics*, 23(2), pp. 27–58.

Anderson, E. (2004) 'Uses of Value Judgments in Science: A General Argument, with Lessons from a Case Study of Feminist Research on Divorce', *Hypatia*, 19(1), pp. 1–24.

Anderson, E. (2006) 'The Epistemology of Democracy', *Episteme*, 3(1–2), pp. 8–22.

Anderson, E. (2010) *The Imperative of Integration*. Princeton, NJ: Princeton University Press.

Anderson, E. (2011) 'Democracy, Public Policy, and Lay Assessments of Scientific Testimony', *Episteme*, 8(2), pp. 144–64.

Anderson, E. (2017) 'Feminist Epistemology and Philosophy of Science', *The Stanford Encyclopedia of Philosophy* (Spring 2017 Edition). Edited by E.N. Zalta. Available at: https://plato.stanford.edu/archives/spr2017/entries/feminism-epistemology/.

Arneson, R.J. (1980) 'Mill versus Paternalism', *Ethics*, 90(4), pp. 470–89.

Ashton, N.A. and McKenna, R. (2020) 'Situating Feminist Epistemology', *Episteme*, 17(1), pp. 28–47.

Audi, R. (1982) 'Believing and Affirming', *Mind*, 91(361), pp. 115–20.

Audi, R. (1997) 'The Place of Testimony in the Fabric of Knowledge and Justification', *American Philosophical Quarterly*, 34(4), pp. 405–22.

Avnur, Y. and Scott-Kakures, D. (2015) 'How Irrelevant Influences Bias Belief', *Philosophical Perspectives*, 29(1), pp. 7–39.

Baehr, J. (2011) *The Inquiring Mind: On Intellectual Virtues and Virtue Epistemology*. New York: Oxford University Press.

Bain, Z. (2018) 'Is There Such a Thing as "White Ignorance" in British Education?', *Ethics and Education*, 13(1), pp. 4–21.

Bak, H.-J. (2001) 'Education and Public Attitudes Toward Science: Implications for the "Deficit Model" of Education and Support for Science and Technology', *Social Science Quarterly*, 82(4), pp. 779–95.

Ballantyne, N. (2019) *Knowing Our Limits*. New York: Oxford University Press.

Ballew, M.T. et al. (2019) 'Climate Change in the American Mind: Data, Tools, and Trends', *Environment: Science and Policy for Sustainable Development*, 61(3), pp. 4–18.

Barnes, B. (1977) *Interests and the Growth of Knowledge*. Abingdon: Routledge and Kegan Paul.

Barnes, B. and Bloor, D. (1982) 'Relativism, Rationalism and the Sociology of Knowledge', in M. Hollis and S. Lukes (eds) *Rationality and Relativism*. Oxford: Blackwell, pp. 21–47.

Baron, J. and Jost, J.T. (2019) 'False Equivalence: Are Liberals and Conservatives in the United States Equally Biased?', *Perspectives on Psychological Science*, 14(2), pp. 292–303.

Battaly, H. (2018) 'Can Closed-Mindedness Be an Intellectual Virtue?', *Royal Institute of Philosophy Supplement*, 84, pp. 23–45.

Battaly, H. (2019) 'Vice Epistemology Has a Responsibility Problem', *Philosophical Issues*, 29(1), pp. 24–36.

Begby, E. (2021) *Prejudice: A Study in Non-Ideal Epistemology*. Oxford: Oxford University Press.

Berenstain, N. (2016) 'Epistemic Exploitation', *Ergo*, 3(22). Available at: https://doi.org/10.3998/ergo.12405314.0003.022.

Berker, S. (2013) 'Epistemic Teleology and the Separateness of Propositions', *Philosophical Review*, 122(3), pp. 337–93.

Berofsky, B. (1983) 'Autonomy', in L.S. Cauman et al. (eds) *How Many Questions? Essays in Honor of Sidney Morgenbesser*. Indianapolis: Hackett, pp. 301–20.

Bilgrami, A. (2006) *Self-Knowledge and Resentment*. Cambridge, MA: Harvard University Press.

Bishop, B. and Cushing, R. (2008) *The Big Sort: Why the Clustering of Like-Minded America is Tearing Us Apart*. Boston, MA: Houghton Mifflin.

Bloor, D. (1976) *Knowledge and Social Imagery*. Chicago, IL: University of Chicago Press.

Boghossian, P. (2006) *Fear of Knowledge: Against Relativism and Constructivism*. Oxford: Oxford University Press.

Bolsen, T. and Druckman, J.N. (2015) 'Counteracting the Politicization of Science', *Journal of Communication*, 65(5), pp. 745–69.

Bolsen, T., Leeper, T.J., and Shapiro, M.A. (2014) 'Doing What Others Do: Norms, Science, and Collective Action on Global Warming', *American Politics Research*, 42(1), pp. 65–89.

Bondy, P. and Carter, J.A. (2019) 'The Superstitious Lawyer's Inference', in P. Bondy and J.A. Carter (eds) *Well-Founded Belief: New Essays on the Epistemic Basing Relation*. Abingdon: Routledge, pp. 125–40.

Bonjour, L. (1980) 'Externalist Theories of Empirical Knowledge', *Midwest Studies in Philosophy*, 5(1), pp. 53–73.

Boorman, E.D. et al. (2013) 'The Behavioral and Neural Mechanisms Underlying the Tracking of Expertise', *Neuron*, 80(6), pp. 1558–71.

Bortolotti, L. (2020) *The Epistemic Innocence of Irrational Beliefs*. Oxford: Oxford University Press.

Bradley, G.W. (1978) 'Self-Serving Biases in the Attribution Process: A Reexamination of the Fact or Fiction Question', *Journal of Personality and Social Psychology*, 36(1), pp. 56–71.

Brown, R. (1986) *Social Psychology*, 2nd edn. New York: Free Press.

Bullock, E.C. (2016) 'Knowing and Not-Knowing for Your Own Good: The Limits of Epistemic Paternalism', *Journal of Applied Philosophy*, 35(2), pp. 433–47.

Bullock, J.G., Gerber, A.S., and Hill, S.J. (2015) 'Partisan Bias in Factual Beliefs about Politics', *Quarterly Journal of Political Science*, 10, pp. 519–78.

Burge, T. (1993) 'Content Preservation', *Philosophical Review*, 102(4), pp. 457–88.

Burnstein, E. and Vinokur, A. (1977) 'Persuasive Argumentation and Social Comparison as Determinants of Attitude Polarization', *Journal of Experimental Social Psychology*, 13(4), pp. 315–32.

Campbell, T.H. and Kay, A.C. (2014) 'Solution Aversion: On the Relation between Ideology and Motivated Disbelief', *Journal of Personality and Social Psychology*, 107(5), pp. 809–24.

Cappelen, H. and Dever, J. (2021) 'On the Uselessness of the Distinction between Ideal and Non-Ideal Theory (at Least in the Philosophy of Language)', in J. Khoo and R. Sterken (eds) *Routledge Companion to Social and Political Philosophy of Language*. Abingdon: Routledge, pp. 91–106.

Carr, J. (forthcoming) 'Why Ideal Epistemology?', *Mind*.

Carter, J.A. (2020) 'Intellectual Autonomy, Epistemic Dependence and Cognitive Enhancement', *Synthese*, 197, pp. 2937–61.

Carter, J.A. (2022) *Autonomous Knowledge: Radical Enhancement, Autonomy, and the Future of Knowing*. Oxford: Oxford University Press.

Carter, J.A. and Littlejohn, C. (2021) *This Is Epistemology*. Hoboken, NJ: Wiley.

Cassam, Q. (2014) *Self-Knowledge for Humans*. Oxford: Oxford University Press.

Cassam, Q. (2016) 'Vice Epistemology', *The Monist*, 99(2), pp. 159–80.

Cassam, Q. (2019) *Vices of the Mind: From the Intellectual to the Political*. Oxford: Oxford University Press.

Coady, C.A.J. (1992) *Testimony: A Philosophical Study*. Oxford: Clarendon Press.

Coady, D. and Chase, J. (eds) (2018) *The Routledge Handbook of Applied Epistemology*. Abingdon: Routledge.

Code, L. (1991) *What Can She Know? Feminist Theory and the Construction of Knowledge*. Ithaca, NY: Cornell University Press.

Cohen, J. (1986) 'An Epistemic Conception of Democracy', *Ethics*, 97(1), pp. 26–38.

Collins, H. (1985) *Changing Order: Replication and Induction in Scientific Practice*. Chicago, IL: University of Chicago Press.

Collins, P.H. (1986) 'Learning from the Outsider Within: The Sociological Significance of Black Feminist Thought', *Social Problems*, 33(6), S14–S32.

Collins, P.H. (2000) *Black Feminist Thought: Knowledge, Consciousness, and the Politics of Empowerment*, 2nd edn. New York: Routledge.

Compton, J. (2013) 'Inoculation Theory', in J.P. Dillard and L. Shen (eds) *The SAGE Handbook of Persuasion: Developments in Theory and Practice*. Thousand Oaks, CA: Sage Publications, pp. 220–36.

Conee, E. and Feldman, R. (1998) 'The Generality Problem for Reliabilism', *Philosophical Studies*, 89(1), pp. 1–29.

Conee, E. and Feldman, R. (2004) *Evidentialism: Essays in Epistemology*. Oxford: Oxford University Press.

Cook, J. (2016) 'Countering Climate Science Denial and Communicating Scientific Consensus', *Oxford Encyclopedia of Climate Change Communication*. Available at: https://doi.org/10.1093/acrefore/9780190228620.013.314.

Cook, J. (2017) 'Understanding and Countering Climate Science Denial', *Journal and Proceedings of the Royal Society of New South Wales*, 150(465/6), pp. 207–19.

Cook, J. and Lewandowsky, S. (2011) *The Debunking Handbook*. St Lucia: University of Queensland.

Cook, J., Lewandowsky, S., and Ecker, U.K.H. (2017) 'Neutralizing Misinformation through Inoculation: Exposing Misleading Argumentation Techniques Reduces their Influence', *PloS One*, 12(5), e0175799.

Cook, J. et al. (2016) 'Consensus on Consensus: A Synthesis of Consensus Estimates on Human-Caused Global Warming', *Environmental Research Letters*, 11(4), 048002.

Cook, J. et al. (2018) *The Consensus Handbook*. Available at: http://www.climatechangecommunication.org/all/consensus-handbook/.

Corner, A. et al. (2015) *The Uncertainty Handbook*. Bristol: University of Bristol Press.

Cvetkovich, G. and Earle, T. (1995) *Social Trust: Toward a Cosmopolitan Society*. Westport, CT: Praeger.

Cvetkovich, G. and Löfstedt, R. (eds) (1999) *Social Trust and the Management of Risk*. Abingdon: Earthscan.

Dahlstrom, M.F. (2014) 'Using Narratives and Storytelling to Communicate Science with Nonexpert Audiences', *Proceedings of the National Academy of Sciences*, 111 (Supplement 4), pp. 13614–20.

Daukas, N. (2019) 'Feminist Virtue Epistemology', in H. Battaly (ed.) *The Routledge Handbook of Virtue Epistemology*. Abingdon: Routledge, pp. 379–91.

Dentith, M.R.X. (2016) 'The Problem of Fake News', *Public Reason*, 8(1–2), pp. 65–79.

DeRose, K. (1995) 'Solving the Skeptical Problem', *Philosophical Review*, 104(1), pp. 1–52.

Dillon, R. (2012) 'Critical Character Theory: Towards a Feminist Perspective on "Vice" (and "Virtue")', in S. Crasnow and A. Superson (eds) *Out from the Shadows: Analytical Feminist Contributions to Traditional Philosophy*. New York: Oxford University Press, pp. 83–114.

DiPaolo, J. and Simpson, R.M. (2016) 'Indoctrination Anxiety and the Etiology of Belief', *Synthese*, 193(10), pp. 3079–98.

Dispensa, J.M. and Brulle, R.J. (2003) 'Media's Social Construction of Environmental Issues: Focus on Global Warming: A Comparative Study', *International Journal of Sociology and Social Policy*, 23(10), pp. 74–105.

Ditto, P.H. et al. (1998) 'Motivated Sensitivity to Preference-Inconsistent Information', *Journal of Personality and Social Psychology*, 75(1), pp. 53–69.

Ditto, P.H. et al. (2019) 'At Least Bias Is Bipartisan: A Meta-Analytic Comparison of Partisan Bias in Liberals and Conservatives', *Perspectives on Psychological Science*, 14(2), pp. 273–91.

Dotson, K. (2011) 'Tracking Epistemic Violence, Tracking Practices of Silencing', *Hypatia*, 26(2), pp. 236–57.

Dotson, K. (2014) 'Conceptualizing Epistemic Oppression', *Social Epistemology*, 28(2), pp. 115–38.

Dotson, K. (2018) 'Accumulating Epistemic Power', *Philosophical Topics*, 46(1), pp. 129–54.

Downing, P. and Ballantyne, J. (2007) *Tipping Point or Turning Point? Social Marketing and Climate Change*. London: Ipsos MORI Social Research Institute.

Dryzek, J.S. and Lo, A.Y. (2015) 'Reason and Rhetoric in Climate Communication', *Environmental Politics*, 24(1), pp. 1–16.

Dworkin, G. (2020) 'Paternalism', *The Stanford Encyclopedia of Philosophy* (Summer 2020 Edition). Edited by E.N. Zalta. Available at: https://plato.stanford.edu/archives/sum2020/entries/paternalism/.

Eddo-Lodge, R. (2018) *Why I'm No Longer Talking to White People About Race*. London: Bloomsbury.

Ellis, J. (2022) 'Motivated Reasoning and the Ethics of Belief', *Philosophy Compass*, 17(6), e12828.

Elzinga, B. (2019) 'A Relational Account of Intellectual Autonomy', *Canadian Journal of Philosophy*, 49(1), pp. 22–47.

Emerson, R.W. (1841) 'Self-Reliance', in *Essays: First Series*. Rahway, NJ: The Mershon Company, pp. 37–80.

Emmet, D. (1994) *The Role of the Unrealisable: A Study in Regulative Ideals*. New York: St. Martin's Press.

Estlund, D. (2019) *Utopophobia: On the Limits (If Any) of Political Philosophy*. Princeton, NJ: Princeton University Press.

Fantl, J. (2018) *The Limitations of the Open Mind*. Oxford: Oxford University Press.

Fantl, J. (2020) 'Review of Knowing Our Limits, by Nathan Ballantyne', *Notre Dame Philosophical Reviews*. Available at: https://ndpr.nd.edu/reviews/knowing-our-limits/.

Fantl, J. and McGrath, M. (2009) *Knowledge in an Uncertain World*. Oxford: Oxford University Press.

Faraji-Rad, A., Samuelsen, B.M., and Warlop, L. (2015) 'On the Persuasiveness of Similar Others: The Role of Mentalizing and the Feeling of Certainty', *Journal of Consumer Research*, 42(3), pp. 458–71.

Feinberg, J. (1986) *Harm to Self*. Oxford: Oxford University Press.

Feldman, R. (1988) 'Epistemic Obligations', *Philosophical Perspectives*, 2, pp. 235–56.

Fine, C. (2010) *Delusions of Gender: How Our Minds, Society, and Neurosexism Create Difference*. New York: Norton.

Flores, C. and Woodard, E. (forthcoming) 'Epistemic Norms on Evidence-Gathering'. *Philosophical Studies*.

Fricker, E. (1994) 'Against Gullibility', in A. Chakrabarti and B.K. Matilal (eds) *Knowing from Words: Western and Indian Philosophical Analysis of Understanding and Testimony*. Dordrecht: Kluwer Academic Publishers, pp. 125–61.

Fricker, E. (2006) 'Testimony and Epistemic Autonomy', in J. Lackey and E. Sosa (eds) *The Epistemology of Testimony*. Oxford: Oxford University Press, pp. 225–53.

Fricker, M. (2007) *Epistemic Injustice: Power and the Ethics of Knowing*. Oxford: Oxford University Press.

Fricker, M. (2010) 'The Relativism of Blame and Williams's Relativism of Distance', *Aristotelian Society Supplementary*, 84(1), pp. 151–77.

Fricker, M. (2016) 'What's the Point of Blame? A Paradigm Based Explanation', *Noûs*, 50(1), pp. 165–83.

Friedman, J. (2020) 'The Epistemic and the Zetetic', *Philosophical Review*, 129(4), pp. 501–36.

Funk, C. and Kennedy, B. (2016) *The Politics of Climate*. Washington, DC: Pew Research Center. Available at: https://www.pewresearch.org/internet/wp-content/uploads/sites/9/2016/10/PS_2016.10.04_Politics-of-Climate_FINAL.pdf.

Gampa, A. et al. (2019) '(Ideo)Logical Reasoning: Ideology Impairs Sound Reasoning', *Social Psychological and Personality Science*, 10(8), pp. 1075–83.

Gardner, G.T. and Stern, P.C. (1996) *Environmental Problems and Human Behavior*. Boston, MA: Allyn & Bacon.

Gerken, M. (2022) *Scientific Testimony: Its Roles in Science and Society*. Oxford: Oxford University Press.

Gettier, E. (1963) 'Is Justified True Belief Knowledge?', *Analysis*, 23(6), pp. 121–3.

Gigerenzer, G. (2000) *Simple Heuristics that Make Us Smart*. Oxford: Oxford University Press.

Ginet, C. (1985) 'Contra Reliabilism', *The Monist*, 68(2), pp. 175–87.

Goldberg, S. (2010) *Relying on Others: An Essay in Epistemology*. Oxford: Oxford University Press.

Goldberg, S. (manuscript) 'Assertion, Silence, and the Norms of Public Reaction'.

Goldenberg, M.J. (2021) *Vaccine Hesitancy: Public Trust, Expertise, and the War on Science*. Pittsburgh, PA: University of Pittsburgh Press.

Goldman, A. (1979) 'What Is Justified Belief?', in G. Pappas and J. Kim (eds) *Justification and Knowledge*. Dordrecht: Springer, pp. 89–104.

Goldman, A. (1986) *Epistemology and Cognition*. Cambridge, MA: Harvard University Press.

Goldman, A. (1999) *Knowledge in a Social World*. Oxford: Oxford University Press.

Goldman, A. (2001) 'Experts: Which Ones Should You Trust?', *Philosophy and Phenomenological Research*, 63(1), pp. 85–110.

Goldman, A. (2010a) 'Systems-Oriented Social Epistemology', *Oxford Studies in Epistemology*, 3, pp. 189–214.

Goldman, A. (2010b) 'Why Social Epistemology Is Real Epistemology', in A. Haddock, A. Millar, and D. Pritchard (eds) *Social Epistemology*. Oxford: Oxford University Press, pp. 1–29.

Goldman, A. (2015) 'Reliabilism, Veritism, and Epistemic Consequentialism', *Episteme*, 12(2), pp. 131–43.

Goldman, A. and O'Connor, C. (2021) 'Social Epistemology', *The Stanford Encyclopedia of Philosophy* (Winter 2021 Edition). Edited by E.N. Zalta. Available at: https://plato.stanford.edu/archives/win2021/entries/epistemology-social.

Graham, P.J. (2012) 'Epistemic Entitlement', *Noûs*, 46(3), pp. 449–82.

Grasswick, H. (2018) 'Feminist Social Epistemology', *The Stanford Encyclopedia of Philosophy* (Fall 2018 Edition). Edited by E.N. Zalta. Available at: https://plato.stanford.edu/archives/fall2018/entries/feminist-social-epistemology/.

Greco, J. (2010) *Achieving Knowledge: A Virtue-Theoretic Account of Epistemic Normativity*. Cambridge: Cambridge University Press.

Grill, K. and Hanna, J. (eds) (2018) *The Routledge Handbook of the Philosophy of Paternalism*. Abingdon: Routledge.

Groll, D. (2012) 'Paternalism, Respect, and the Will', *Ethics*, 122(4), pp. 692–720.

Hamilton, L.C. (2011) 'Education, Politics and Opinions about Climate Change Evidence for Interaction Effects', *Climatic Change*, 104(2), pp. 231–42.

Hamilton, L.C. et al. (2015) 'Tracking Public Beliefs about Anthropogenic Climate Change', *PLoS One*, 10(9), e0138208.

Hannon, M. (2020) 'Empathetic Understanding and Deliberative Democracy', *Philosophy and Phenomenological Research*, 101(3), pp. 591–611.

Harding, S. (1991) *Whose Science? Whose Knowledge? Thinking from Women's Lives*. Ithaca, NY: Cornell University Press.

Harding, S. (1995) '"Strong Objectivity": A Response to the New Objectivity Question', *Synthese*, 104(3), pp. 331–49.

Hardisty, D.J., Johnson, E.J., and Weber, E.U. (2010) 'A Dirty Word or a Dirty World? Attribute Framing, Political Affiliation, and Query Theory', *Psychological Science*, 21(1), pp. 86–92.

Harman, G.H. (1970) 'Knowledge, Reasons, and Causes', *Journal of Philosophy*, 67(21), pp. 841–55.

Hartsock, N. (1983) 'The Feminist Standpoint: Developing the Ground for a Specifically Feminist Historical Materialism', in S. Harding and M.B. Hintikka (eds) *Discovering Reality: Feminist Perspectives on Epistemology, Metaphysics, Methodology, and the Philosophy of Science*. Dordrecht: Reidel, pp. 283–310.

Hausman, D.M. and Welch, B. (2010) 'Debate: To Nudge or Not to Nudge', *Journal of Political Philosophy*, 18(1), pp. 123–136.

Higgins, L. (2016) 'More Michigan Parents Willing to Vaccinate Kids', *Detroit Free Press*. Available at: https://eu.freep.com/story/news/education/2016/01/28/immunization-waivers-plummet-40-michigan/79427752/.

Hodson, G. and Busseri, M. (2012) 'Bright Minds and Dark Attitudes: Lower Cognitive Ability Predicts Greater Prejudice Through Right-Wing Ideology and Low Intergroup Contact', *Psychological Science*, 23(2), pp. 187–95.

Hookway, C. (2003) 'How to Be a Virtue Epistemologist', in L. Zagzebski and M. DePaul (eds) *Intellectual Virtue: Perspectives from Ethics and Epistemology*. Oxford: Oxford University Press, pp. 183–202.

Hornsey, M.J. et al. (2016) 'Meta-Analyses of the Determinants and Outcomes of Belief in Climate Change', *Nature Climate Change*, 6(6), pp. 622–6.

Howe, L.C. and Krosnick, J.A. (2017) 'Attitude Strength', *Annual Review of Psychology*, 68, pp. 327–51.

Huemer, M. (2016) 'Why People Are Irrational about Politics', in G. Brennan, M.C. Munger, and G. Sayre-McCord (eds) *Philosophy, Politics and Economics: An Anthology*. New York: Oxford University Press, pp. 456–67.

Hughes, N. (forthcoming) 'Epistemic Feedback Loops (Or: How Not to Get Evidence)', *Philosophy and Phenomenological Research*.

Hume, D. (2007) *An Enquiry Concerning Human Understanding*, ed. P. Millican. Oxford: Oxford University Press.

Hutchison, K., Mackenzie, C., and Oshana, M. (eds) (2018) *Social Dimensions of Moral Responsibility*. New York: Oxford University Press.

Ipsos MORI (2014) *Public Attitudes to Science*. Available at: https://www.britishscience association.org/public-attitudes-to-science-survey.

Ivanov, B. et al. (2015) 'The General Content of Postinoculation Talk: Recalled Issue-Specific Conversations Following Inoculation Treatments', *Western Journal of Communication*, 79(2), pp. 218–38.

Joshi, H. (2020) 'What Are the Chances You're Right about Everything? An Epistemic Challenge for Modern Partisanship', *Politics, Philosophy and Economics*, 19(1), pp. 36–61.

Joshi, H. (2021) *Why It's OK to Speak Your Mind*. New York: Routledge.

Jost, J.T., Hennes, E.P., and Lavine, H. (2013) '"Hot" Political Cognition: Its Self-, Group-, and System-Serving Purposes', in D.E. Carlston (ed.) *The Oxford Handbook of Social Cognition*. New York: Oxford University Press, pp. 851–75.

Kahan, D. (2010) 'Fixing the Communications Failure', *Nature*, 463, pp. 296–7.

Kahan, D. (2013) 'Ideology, Motivated Reasoning, and Cognitive Reflection', *Judgment and Decision Making*, 8(4), pp. 407–24.

Kahan, D. (2014) 'Making Climate-Science Communication Evidence-Based—All the Way Down', in M.T. Boykoff and D.A. Crow (eds) *Culture, Politics and Climate Change: How Information Shapes Our Common Future*. New York: Routledge, pp. 203–20.

Kahan, D. (2016a) 'The Politically Motivated Reasoning Paradigm, Part 1: What Politically Motivated Reasoning Is and How to Measure It', in R.A. Scott and M.C. Buchman (eds) *Emerging Trends in the Social and Behavioral Sciences: An Interdisciplinary, Searchable, and Linkable Resource*. New York: Wiley, pp. 1–16.

Kahan, D. (2016b) 'The Politically Motivated Reasoning Paradigm, Part 2: Unanswered Questions', in R.A. Scott and M.C. Buchman (eds) *Emerging Trends in the Social and Behavioral Sciences: An Interdisciplinary, Searchable, and Linkable Resource*. New York: Wiley, pp. 1–15.

Kahan, D., Hoffman, D., et al. (2012) '"They Saw a Protest": Cognitive Illiberalism and the Speech-Conduct Distinction', *Stanford Law Review*, 64(4), pp. 851–906.

Kahan, D., Jenkins-Smith, H., and Braman, D. (2011) 'Cultural Cognition of Scientific Consensus', *Journal of Risk Research*, 14(2), pp. 147–74.

Kahan, D., Peters, E., et al. (2012) 'The Polarizing Impact of Science Literacy and Numeracy on Perceived Climate Change Risks', *Nature Climate Change*, 2, pp. 732–5.

Kahan, D. et al. (2009) 'Cultural Cognition of the Risks and Benefits of Nanotechnology', *Nature Nanotechnology*, 4(2), pp. 87–91.

Kahan, D. et al. (2010) 'Who Fears the HPV Vaccine, Who Doesn't, and Why? An Experimental Study of the Mechanisms of Cultural Cognition', *Law and Human Behavior*, 34(6), pp. 501–16.

Kahan, D. et al. (2011) 'The Tragedy of the Risk-Perception Commons: Culture Conflict, Rationality Conflict, and Climate Change', Temple University Legal Studies Research Paper No. 2011–26. Available at: https://ssrn.com/abstract=1871503.

Kahan, D. et al. (2015) 'Geoengineering and Climate Change Polarization: Testing a Two-Channel Model of Science Communication', *Annals of American Academy of Political and Social Science*, 658, pp. 193–222.

Kahan, D. et al. (2016) '"Ideology" or "Situation Sense"? An Experimental Investigation of Motivated Reasoning and Professional Judgment', *University of Pennsylvania Law Review*, 164(349), pp. 349–438.

Kahan, D. et al. (2017) 'Science Curiosity and Political Information Processing', *Political Psychology*, 38(S1), pp. 179–99.

Kanazawa, S. (2010) 'Why Liberals and Atheists Are More Intelligent', *Social Psychology Quarterly*, 73(1), pp. 33–57.

Kant, I. (1959) 'What Is Enlightenment?', in L.W. Beck (trans.) *Foundations of the Metaphysics of Morals and What Is Enlightenment*. New York: Liberal Arts Press.

Keller, E.F. (1985) *Reflections on Gender and Science*. New Haven, CT: Yale University Press.

Kelly, T. (2003) 'Epistemic Rationality as Instrumental Rationality: A Critique', *Philosophy and Phenomenological Research*, 66(3), pp. 612–40.

Kelly, T. (2008) 'Disagreement, Dogmatism, and Belief Polarization', *Journal of Philosophy*, 105(10), pp. 611–33.

Keren, A. (2013) 'Kitcher on Well-Ordered Science: Should Science Be Measured against the Outcomes of Ideal Democratic Deliberation?', *Theoria: An International Journal for Theory, History and Foundations of Science*, 28(2), pp. 233–44.

Kim, J. (1988) 'What Is "Naturalized Epistemology?"', *Philosophical Perspectives*, 2, pp. 381–405.

Kitcher, P. (2001) *Science, Truth, and Democracy*. Oxford: Oxford University Press.

Kitcher, P. (2011) *Science in a Democratic Society*. Buffalo, NY: Prometheus Books.

Korcz, K. (2000) 'The Causal-Doxastic Theory of the Basing Relation', *Canadian Journal of Philosophy*, 30(4), pp. 525–50.

Kornblith, H. (1993) 'Epistemic Normativity', *Synthese*, 94(3), pp. 357–76.

Kornblith, H. (1999) 'Distrusting Reason', *Midwest Studies in Philosophy*, 23(1), pp. 181–96.

Kornblith, H. (2012) *On Reflection*. Oxford: Oxford University Press.

Kornblith, H. (2019) *Second Thoughts and the Epistemological Enterprise*. Cambridge: Cambridge University Press.

Kunda, Z. (1987) 'Motivated Inference: Self-Serving Generation and Evaluation of Causal Theories', *Journal of Personality and Social Psychology*, 53(4), pp. 636–47.

Kunda, Z. (1990) 'The Case for Motivated Reasoning', *Psychological Bulletin*, 108(3), pp. 480–98.

Kusch, M. (2010) 'Social Epistemology', in S. Bernecker and D. Pritchard (eds) *Routledge Companion to Epistemology*. New York: Routledge, pp. 873–84.

Kvanvig, J. (2003) 'Propositionalism and the Perspectival Character of Justification', *American Philosophical Quarterly*, 40(1), pp. 3–17.

Lacey, H. (1999) *Is Science Value Free? Values and Scientific Understanding*. Abingdon: Routledge.

Lackey, J. (2007) 'Norms of Assertion', *Noûs*, 41(4), pp. 594–626.

Lackey, J. (2018) 'Silence and Objecting', in C.R. Johnson (ed.) *Voicing Dissent: The Ethics and Epistemology of Making Disagreement Public*. New York: Routledge, pp. 82–96.

Lackey, J. (2020) 'The Duty to Object', *Philosophy and Phenomenological Research*, 101(1), pp. 35–60.

Landemore, H. (2013) 'Deliberation, Cognitive Diversity, and Democratic Inclusiveness: An Epistemic Argument for the Random Selection of Representatives', *Synthese*, 190(7), pp. 1209–31.

Lasonen-Aarnio, M. (2010) 'Unreasonable Knowledge', *Philosophical Perspectives*, 24(1), pp. 1–21.

Levy, N. (2019) 'Due Deference to Denialism: Explaining Ordinary People's Rejection of Established Scientific Findings', *Synthese*, 196(1), pp. 313–27.

Levy, N. (2021) *Bad Beliefs: Why They Happen to Good People*. Oxford: Oxford University Press.

Lewandowsky, S., Gignac, G.E., and Vaughan, S. (2013) 'The Pivotal Role of Perceived Scientific Consensus in Acceptance of Science', *Nature Climate Change*, 3(4), pp. 399–404.

Lewandowsky, S. and Oberauer, K. (2016) 'Motivated Rejection of Science', *Current Directions in Psychological Science*, 25(4), pp. 217–22.

Lewandowsky, S. et al. (2012) 'Misinformation and Its Correction: Continued Influence and Successful Debiasing', *Psychological Science in the Public Interest*, 13(3), pp. 106–31.

Lin, H. (2022) 'Bayesian Epistemology', in E.N. Zalta (ed.) *Stanford Encyclopedia of Philosophy* (Summer 2022 Edition). Available at: https://plato.stanford.edu/archives/sum2022/entries/epistemology-bayesian/.

Longino, H. (1997) 'Feminist Epistemology as a Local Epistemology', *Aristotelian Society Supplementary*, 71(1), pp. 19–36.

Lord, C.G., Ross, L., and Lepper, M.R. (1979) 'Biased Assimilation and Attitude Polarization: The Effects of Prior Theories on Subsequently Considered Evidence', *Journal of Personality and Social Psychology*, 37(11), pp. 2098–109.

Lynch, M.P. (2016) *Internet of Us: Knowing More and Understanding Less in the Age of Big Data*. New York: W.W. Norton.

Marks, J. et al. (2019) 'Epistemic Spillovers: Learning Others' Political Views Reduces the Ability to Assess and Use Their Expertise in Nonpolitical Domains', *Cognition*, 188, pp. 74–84.

Martín, A. (2021) 'What Is White Ignorance?', *The Philosophical Quarterly*, 71(4), pp. 864–85.

MacInnis, B. et al. (2015) 'The American Public's Preference for Preparation for the Possible Effects of Global Warming: Impact of Communication Strategies', *Climatic Change*, 128(1–2), pp. 17–33.

MacLeod, C. (2021) 'Mill on the Liberty of Thought and Discussion', in A. Stone and F. Schauer (eds) *The Oxford Handbook of the Freedom of Speech*. Oxford: Oxford University Press, pp. 3–19.

McCain, K. (2012) 'The Interventionist Account of Causation and the Basing Relation', *Philosophical Studies*, 159(3), pp. 357–82.

McCormick, M.S. (2020) 'Believing Badly: Doxastic Duties Are Not Epistemic Duties', in K. McCain and S. Stapleford (eds) *Epistemic Duties: New Arguments, New Angles*. Abingdon: Routledge, pp. 29–43.

McGlynn, A. (2019) 'Redrawing the Map: Medina on Epistemic Vices and Skepticism', *International Journal for the Study of Skepticism*, 9(3), pp. 261–83.

McKenna, R. (2020) 'Pragmatic Encroachment and Feminist Epistemology', in N.A. Ashton et al. (eds) *Social Epistemology and Epistemic Relativism*. Abingdon: Routledge, pp. 103–21.

McNeil, B.J. et al. (1982) 'On the Elicitation of Preferences for Alternative Therapies', *The New England Journal of Medicine*, 306(21), pp. 1259–62.

Medina, J. (2012) *The Epistemology of Resistance: Gender and Racial Oppression, Epistemic Injustice, and Resistant Imaginations*. New York: Oxford University Press.

Meehan, D. (2020) 'Epistemic Vice and Epistemic Nudging: A Solution', in G. Axtell and A. Bernal (eds) *Epistemic Paternalism: Conceptions, Justifications and Implications*. London: Rowman & Littlefield, pp. 247–59.

Mill, J.S. (2011) *On Liberty*. Cambridge: Cambridge University Press.

Millar, A. (2019) *Knowing by Perceiving*. Oxford: Oxford University Press.

Miller, B. (2013) 'When Is Consensus Knowledge Based? Distinguishing Shared Knowledge from Mere Agreement', *Synthese*, 190(7), pp. 1293–316.

Mills, C. (1997) *The Racial Contract*. Ithaca, NY: Cornell University Press.

Mills, C. (2005) '"Ideal Theory" as Ideology', *Hypatia*, 20(3), pp. 165–84.

Mills, C. (2007) 'White Ignorance', in N. Tuana and S. Sullivan (eds) *Race and Epistemologies of Ignorance*. Albany, NY: State University of New York Press, pp. 13–38.

Mitchell, G. (2005) 'Libertarian Paternalism is an Oxymoron', *Northwestern University Law Review*, 99(3), pp. 1245–77.

Mogensen, A.L. (2016) 'Contingency Anxiety and the Epistemology of Disagreement', *Pacific Philosophical Quarterly*, 97(4), pp. 590–611.

Molden, D.C. and Higgins, E.T. (2012) 'Motivated Thinking', in K.J. Holyoak and R.G. Morrison (eds) *The Oxford Handbook of Thinking and Reasoning*. Oxford: Oxford University Press, pp. 390–409.

Moser, P. (1989) *Knowledge and Evidence*. Cambridge: Cambridge University Press.

Moser, S.C. and Dilling, L. (2011) 'Communicating Climate Change: Closing the Science–Action Gap', in J.S. Dryzek, R.B. Norgaard, and D. Schlosberg (eds) *The Oxford Handbook of Climate Change and Society*. Oxford: Oxford University Press, pp. 161–74.

National Research Council Board on Radioactive Waste Management (1990) *Rethinking High-Level Radioactive Waste Disposal: A Position Statement of the Board on Radioactive Waste Management*. Washington, DC: National Academies Press.

National Research Council Committee on Analysis of Global Change Assessments (2007) *Analysis of Global Change Assessments: Lessons Learned*. Washington, DC: National Academies Press.

National Research Council Committee to Improve Research Information and Data on Firearms (2004) *Firearms and Violence: A Critical Review*. Washington, DC: National Academies Press.

Navin, M.C. and Largent, M.A. (2017) 'Improving Nonmedical Vaccine Exemption Policies: Three Case Studies', *Public Health Ethics*, 10(3), pp. 225–34.

Newport, F. (2010) 'Americans' Global Warming Concerns Continue to Drop', *Gallup*. Available at: https://news.gallup.com/poll/126560/Americans-Global-Warming-Concerns-Continue-Drop.aspx (accessed 13 July 2020).

Nguyen, C.T. (2020) 'Cognitive Islands and Runaway Echo Chambers: Problems for Epistemic Dependence on Experts', *Synthese*, 197(7), pp. 2803–21.

Nyhan, B. and Reifler, J. (2010) 'When Corrections Fail', *Political Behavior*, 32(2), pp. 303–30.

Pasnau, R. (2013) 'Epistemology Idealized', *Mind*, 122(488), pp. 987–1021.

Pasnau, R. (2017) *After Certainty: A History of Our Epistemic Ideals and Illusions*. Oxford: Oxford University Press.

Pfau, M. (1995) 'Designing Messages for Behavioral Inoculation', in E.H. Maibach and R.L. Parrott (eds) *Designing Health Messages: Approaches from Communication Theory and Public Health Practice*. Thousand Oaks, CA: Sage Publications, pp. 99–113.

Pfau, M. and Burgoon, M. (1988) 'Inoculation in Political Campaign Communication', *Human Communication Research*, 15(1), pp. 91–111.

Piovarchy, A. (2021) 'Responsibility for Testimonial Injustice', *Philosophical Studies*, 178(2), pp. 597–615.

Pollock, J.L. and Cruz, J. (1999) *Contemporary Theories of Knowledge*. Lanham, MD: Rowman & Littlefield.

Prior, M., Sood, G., and Khanna, K. (2015) 'You Cannot Be Serious: The Impact of Accuracy Incentives on Partisan Bias in Reports of Economic Perceptions', *Quarterly Journal of Political Science*, 10, pp. 489–518.

Pritchard, D. (2002) 'Resurrecting the Moorean Response to the Sceptic', *International Journal of Philosophical Studies*, 10(3), pp. 283–307.

Pritchard, D. (2010) 'Relevant Alternatives, Perceptual Knowledge and Discrimination', *Noûs*, 44(2), pp. 245–68.

Pritchard, D. (2012) *Epistemological Disjunctivism*. Oxford: Oxford University Press.

Quine, W.V.O. (1986) 'Reply to Morton White', in L. Hahn and P. Schilpp (eds) *The Philosophy of W. V. Quine*. Chicago, IL: Open Court, pp. 663–5.

Rawls, J. (1971) *A Theory of Justice*. Cambridge, MA: Harvard University Press.

Redlawsk, D.P. (2002) 'Hot Cognition or Cool Consideration? Testing the Effects of Motivated Reasoning on Political Decision Making', *The Journal of Politics*, 64(4), pp. 1021–44.

Rescher, N. (1987) *Ethical Idealism: An Inquiry into the Nature and Function of Ideals*. Berkeley, CA: University of California Press.

Riley, E. (2017) 'The Beneficent Nudge Program and Epistemic Injustice', *Ethical Theory and Moral Practice*, 20(3), pp. 597–616.

Riley, J. (2018) 'Mill's Absolute Ban on Paternalism', in K. Grill and J. Hanna (eds) *The Routledge Handbook of the Philosophy of Paternalism*. Abingdon: Routledge, pp. 153–69.

Rini, R. (2017) 'Fake News and Partisan Epistemology', *Kennedy Institute of Ethics Journal*, 27(S2), pp. 43–64.

Roberts, R.C. and Wood, W.J. (2007) *Intellectual Virtues: An Essay in Regulative Epistemology*. Oxford: Clarendon Press.

Rolin, K. (2002) 'Gender and Trust in Science', *Hypatia*, 17(4), pp. 95–118.

Samuels, R., Stich, S., and Bishop, M. (2002) 'Ending the Rationality Wars: How to Make Disputes about Human Rationality Disappear', in R. Elio (ed.) *Common Sense, Reasoning and Rationality*. Oxford: Oxford University Press, pp. 236–68.

Schilbach, L. et al. (2013) 'To You I Am Listening: Perceived Competence of Advisors Influences Judgment and Decision-Making Via Recruitment of the Amygdala', *Social Neuroscience*, 8(3), pp. 189–202.

Scoccia, D. (2018) 'The Concept of Paternalism', in K. Grill and J. Hanna (eds) *The Routledge Handbook of the Philosophy of Paternalism*. Abingdon: Routledge, pp. 11–23.

Seifert, C.M. (2002) 'The Continued Influence of Misinformation in Memory: What Makes a Correction Effective?', in B.H. Ross (ed.) *The Psychology of Learning and Motivation: Advances in Research and Theory*. Cambridge: Academic Press, pp. 265–92.

Shiffrin, S.V. (2000) 'Paternalism, Unconscionability Doctrine, and Accommodation', *Philosophy and Public Affairs*, 29(3), pp. 205–50.

Simion, M., Kelp, C., and Ghijsen, H. (2016) 'Norms of Belief', *Philosophical Issues*, 26(1), pp. 374–92.

Skurnik, I., Yoon, C., and Schwarz, N. (2005) 'How Warnings About False Claims Become Recommendations', *Journal of Consumer Research*, 31(4), pp. 713–24.

Sosa, E. (1999) 'How to Defeat Opposition to Moore', *Philosophical Perspectives*, 13(s13), pp. 137–49.

Sosa, E. (2003) 'The Place of Truth in Epistemology', in M. DePaul and L. Zagzebski (eds) *Intellectual Virtues: Perspectives from Ethics and Epistemology*. Oxford: Oxford University Press, pp. 155–80.

Sosa, E. (2007) *A Virtue Epistemology: Apt Belief and Reflective Knowledge, Volume I*. Oxford: Oxford University Press.

Sosa, E. (2009) *Reflective Knowledge: Apt Belief and Reflective Knowledge, Volume II*. Oxford: Oxford University Press.

Sosa, E. (2015) *Judgment and Agency*. Oxford: Oxford University Press.

Srinivasan, A. (2015) 'The Archimedean Urge', *Philosophical Perspectives*, 29(1), pp. 325–62.

Srinivasan, A. (2020) 'Radical Externalism', *Philosophical Review*, 129(3), pp. 395–431.

Sturgis, P. and Allum, N. (2004) 'Science in Society: Re-Evaluating the Deficit Model of Public Attitudes', *Public Understanding of Science*, 13(1), pp. 55–74.

Sturm, T. (2012) 'The "Rationality Wars" in Psychology: Where They Are and Where They Could Go', *Inquiry*, 55(1), pp. 66–81.

Swain, M. (1981) *Reasons and Knowledge*. Ithaca, NY: Cornell University Press.

Taber, C.S. and Lodge, M. (2006) 'Motivated Skepticism in the Evaluation of Political Beliefs', *American Journal of Political Science*, 50(3), pp. 755–69.

Talisse, R.B. (2019) *Overdoing Democracy: Why We Must Put Politics in Its Place*. New York: Oxford University Press.

Tanesini, A. (2016) ' "Calm Down, Dear": Intellectual Arrogance, Silencing and Ignorance', *Aristotelian Society Supplementary Volume*, 90(1), pp. 71–92.

Tanesini, A. (2018) 'Collective Amnesia and Epistemic Injustice', in J.A. Carter et al. (eds) *Socially Extended Epistemology*. Oxford: Oxford University Press, pp. 195–219.

Tanesini, A. (2021) *The Mismeasure of the Self: A Study in Vice Epistemology*. Oxford: Oxford University Press.

Tanesini, A. (2022) 'Intellectual Autonomy and Its Vices', in J. Matheson and K. Lougheed (eds) *Epistemic Autonomy*. Abingdon: Routledge, pp. 173–94.

Tessman, L. (2005) *Burdened Virtues: Virtue Ethics for Liberatory Struggles*. Oxford: Oxford University Press.

Thaler, R.H. and Sunstein, C.R. (2003) 'Libertarian Paternalism', *The American Economic Review*, 93(2), pp. 175–9.

Thaler, R.H. and Sunstein, C.R. (2008) *Nudge: Improving Decisions about Health, Wealth, and Happiness*. New Haven, CT: Yale University Press.

Thaler, R.H., Sunstein, C.R., and Balz, J.P. (2013) 'Choice Architecture', in E. Shafir (ed.) *The Behavioral Foundations of Public Policy*. Princeton, NJ: Princeton University Press, pp. 428–39.

Toole, B. (2019) 'From Standpoint Epistemology to Epistemic Oppression', *Hypatia*, 34(4), pp. 598–618.

Toole, B. (2022) 'Demarginalizing Standpoint Epistemology', *Episteme*, 19(1), pp. 47–65.

Tranter, B. and Booth, K. (2015) 'Scepticism in a Changing Climate: A Cross-National Study', *Global Environmental Change*, 33, pp. 154–64.

Tsai, G. (2014) 'Rational Persuasion as Paternalism', *Philosophy and Public Affairs*, 42(1), pp. 78–112.

Turri, J. (2011) 'Believing for a Reason', *Erkenntnis*, 74(3), pp. 383–97.

Valentini, L. (2012) 'Ideal vs. Non-ideal Theory: A Conceptual Map', *Philosophy Compass*, 7(9), pp. 654–64.

van der Linden, S. et al. (2014) 'How to Communicate the Scientific Consensus on Climate Change: Plain Facts, Pie Charts or Metaphors?', *Climatic Change*, 126(1–2), pp. 255–62.

van der Linden, S. et al. (2017) 'Inoculating the Public Against Misinformation about Climate Change', *Global Challenges*, 1(2), 1600008.

Vargas, M. (2018) 'The Social Constitution of Agency and Responsibility: Oppression, Politics, and Moral Ecology', in K. Hutchison, C. Mackenzie, and M. Oshana (eds) *Social Dimensions of Moral Responsibility*. Oxford: Oxford University Press, pp. 111–34.

Vavova, K. (2018) 'Irrelevant Influences', *Philosophy and Phenomenological Research*, 96(1), pp. 134–52.

Watson, G. (1996) 'Two Faces of Responsibility', *Philosophical Topics*, 24(2), pp. 227–48.

Weisberg, M. (2007) 'Three Kinds of Idealization', *Journal of Philosophy*, 104(12), pp. 639–59.

White, R. (2010) 'You Just Believe that Because...', *Philosophical Perspectives*, 24(1), pp. 573–615.

Williamson, T. (2000) *Knowledge and Its Limits*. Oxford: Oxford University Press.

Wolterstorff, N. (1996) *John Locke and the Ethics of Belief*. Cambridge: Cambridge University Press.

Wrenn, C.B. (2007) 'Why There Are No Epistemic Duties', *Dialogue: The Canadian Philosophical Review*, 46(1), pp. 115–36.

Yancy, G. and Mills, C. (2014) 'Lost in Rawlsland', *The New York Times*. Available at: https://opinionator.blogs.nytimes.com/2014/11/16/lost-in-rawlsland/.

Young, I.M. (2006) 'Responsibility and Global Justice: A Social Connection Model', *Social Philosophy and Policy*, 23(1), pp. 102–30.

Young, I.M. (2011) *Responsibility for Justice*. Oxford: Oxford University Press.

Zagzebski, L. (1996) *Virtues of the Mind: An Inquiry into the Nature of Virtue and the Ethical Foundations of Knowledge*. Cambridge: Cambridge University Press.

Zagzebski, L. (2013) 'Intellectual Autonomy', *Philosophical Issues*, 23(1), pp. 244–61.

Zheng, R. (2018) 'What Is My Role in Changing the System? A New Model of Responsibility for Structural Injustice', *Ethical Theory and Moral Practice*, 21(4), pp. 869–85.

Zheng, R. (2021) 'Moral Criticism and Structural Injustice', *Mind*, 130(518), pp. 503–35.

Index

For the benefit of digital users, indexed terms that span two pages (e.g., 52–53) may, on occasion, appear on only one of those pages.

abstraction 24–5, 39
active ignorance 138–42, 146–7, 151–3
    (*see also* epistemic injustice; responsibility; white ignorance)
admiration 115
Ahlstrom–Vij, Kristoffer 71–2
Alfano, Mark 14n.3, 32
Anderson, Elizabeth 10, 13, 15–16, 37–8, 50–6, 58–60, 62, 105, 137
applied epistemology 57
Ashton, Natalie 13
autonomy 71, 73–4, 82–3 (*see also* paternalism)

Baehr, Jason 134
balance in journalism 53, 55
Ballantyne, Nathan 14n.3, 40n.4, 128–9
Barnes, Barry 11
Battaly, Heather 113
basing
    causal accounts of basing 171–3
    doxastic accounts of basing 170, 172
    inferential accounts of basing 170, 172–3
    proper basing 167–9, 171–3
    see also propositional vs. doxastic justification
Bayesian epistemology 39
Begby, Endre 21n.2, 166
beneficent nudge programme, the 74–7
    (*see also* nudges)
bias 9–10, 32, 38, 55–6 (*see also* motivated reasoning; prejudice)
blame and blameworthiness 110, 115, 143–6 (*see also* responsibility)
Bloor, David 11
Bortolotti, Lisa 21
brainwashing 61–2
Bullock, John G. 99

Cappelen, Herman 42–3
Carr, Jennifer 11–12
Carter, J. Adam 89–90
Cassam, Quassim 13–14, 16–17, 32, 80, 110–12, 114–15, 117, 120–1
choice architecture 72–4, 93–5

closed-mindedness 79–80, 110–11, 113, 139
    (*see also* intellectual vices; vice epistemology)
Code, Lorraine 8, 13
cognitive enhancement 89–90, 94
Collins, Harry 11
consequential false beliefs, problem of 61–2, 85–6
conservative attitudes towards science 53–4, 66–9, 159–62, 165 (*see also* liberal attitudes towards science; motivated reasoning; political ideology)
colour blindness 26–7
credibility 122–3
critical reflection 75–6, 80–1

Daukas, Nancy 135–7
debate 29, 42, 120, 122–4 (*see also* engagement)
debunking 69–70
debunking arguments 174–5
defeat 117n.3, 128–9, 174–5
democratic legitimacy 50–1, 53, 60, 62
Dever, Josh 42–3
dialogue (*see* debate)
Dilling, Lisa 67
Dillon, Robin 142–4, 146–8
disappointment 115
distrust 49, 54, 82–5
division of cognitive labour, the 150–2
dogmatism (*see* closed–mindedness)
Dotson, Kristie 117–20

education 38, 67
Elzinga, Benjamin 92
Emmet, Dorothy 10n.2, 98–9
empathy 136, 155
engagement
    benefits and costs of engagement 107–8, 113–14
    engaging with challenges to our beliefs 16–17, 29, 42, 105–7, 120–1
    see also debate; open-mindedness

epistemic agency 82–3, 135, 149–50, 154
  (*see also* epistemic responsibility)
epistemic consequentialism 105n.1, 108–9
epistemic dependence 89, 91
epistemic environment 15–16, 33, 50, 55–6,
  63–4, 77–8, 168–9 (*see also* idealization;
  institutional epistemology)
epistemic exclusion (*see* epistemic oppression)
epistemic externalism 124–5, 127
  (*see also* reliabilism)
epistemic injustice 75–6, 142
  responsibility for epistemic injustice 145–8,
  151–3
  testimonial injustice 144
  see also active ignorance; prejudice;
  responsibility; white ignorance
epistemic norms (*see* epistemic
  obligations; inquiry epistemology; norms of
  inquiry)
epistemic obligations 7–8, 28–9, 105, 121, 123–4
  (*see also* inquiry epistemology; norms of
  inquiry)
epistemic oppression 117–19, 123–4
  (*see also* epistemic injustice)
epistemic paternalism 70–2, 100–1
  (*see also* intellectual autonomy; nudges;
  paternalism)
epistemic responsibility 51–2, 155–6
  (*see also* epistemic agency; responsibility)
epistemic vices (*see* intellectual vices)
epistemology of disagreement 8, 40
epistemology of testimony 8, 32–3, 116–17
Estlund, David 26
evidentialism 165–7
expert deference 51–2, 55, 97–8, 159
  (*see also* identifying experts, the
  problem of)
expert testimony 46–8

fallacy of approximation, the 26
fallibilism 177–8
Fantl, Jeremy 125–9
feasibility constraints 23, 30
feminist epistemology 4, 8–10, 12–13
  feminist standpoint theory 6, 32, 138
  see also feminist virtue ethics; liberatory
  epistemology; social situation)
feminist virtue ethics 142–4 (*see also* feminist
  epistemology)
'follow the experts' 159
framing 69, 72–4, 99 (*see also* science marketing)
freedom of expression 107–8
Fricker, Elizabeth 88–9
Fricker, Miranda 76, 115, 122, 144–6

full vs. partial compliance theory 22–3, 27–9, 121
  (*see also* idealization; ideal vs. non-ideal
  theory; ideal epistemology)

Gettier, Edmund 2
Gerken, Mikkel 63–4
Ghijsen, Harmen 36
Gigerenzer, Gerd 168
Goldenberg, Maya 49
Goldberg, Sandy 130–1
Goldman, Alvin 6–7, 46–50, 56–8, 165
Gricean norms 130

Hausman, Daniel 73–4
hermeneutical gaps 148

idealization
  minimalist idealization 39–40
  problems with idealization 3, 20, 25–7
  types of idealization 11–12, 25, 31–5
  see also ideal epistemology; ideal and non-ideal
  theory
ideal epistemology
  epistemic ideals, the theory of 30, 37, 177–8
  problems with ideal epistemology 10, 34–5,
  39–42
  see also idealization; ideal and non-ideal
  theory
ideal and non-ideal theory
  demandingness 26–7, 155, 179
  guidance 26–7, 91
  ideal and non-ideal theory in philosophy of
  language 42–3
  ideal and non-ideal theory in political
  philosophy 22–8
  see also idealization; ideal epistemology
identifying experts, the problem of 45–6
  (*see also* expert deference)
information environment (*see* epistemic
  environment)
injustice 19, 37, 152 (*see also* epistemic injustice)
inoculation theory 70, 74
inquiry epistemology 13–15, 24, 28–31, 34–5, 37,
  40–1, 108–9 (*see also* norms of inquiry)
institutional epistemology 6–7, 55, 63–5
intellectual arrogance 119–20, 139
  (*see also* intellectual vices)
intellectual autonomy
  intellectual autonomy as a capacity to reason
  critically 76–8
  modest vs. radical conceptions of intellectual
  autonomy 88–9
  relational conceptions of intellectual
  autonomy 92

value of intellectual autonomy 90, 95–6,
    98–102
intellectual character 79–80, 110, 134,
    143–4, 146
intellectual diligence 139
intellectual humility 89, 139
intellectual virtues 133–4, 136–7, 139, 143
    (*see also* intellectual autonomy; intellectual
    diligence; intellectual humility;
    open–mindedness; virtue epistemology)
intellectual vices 78–81, 110–11, 139–41, 143–4,
    149–50 (*see also* closed–mindedness;
    intellectual arrogance; responsibility; vice
    epistemology)

Joshi, Hrishikesh 163n.2
justice, the theory of 22–3

Kahan, Dan 54, 77, 97–8, 102–3, 158–60,
    162–3
Kelp, Christoph 36
Keren, Arnon 33
Kim, Jaegwon 36
Kitcher, Philip 7, 33, 64
Kornblith, Hilary 21, 32, 177
knowledge
    knowledge, production of 6–7, 64–5,
        117–20
    knowledge, theory of 8, 13–14, 19n.1, 30,
        37–8, 40–1, 110–12, 134, 175, 177–9
    knowledge, value of 137–8
    self–knowledge 149–50
knowledge-action principle, the 127
knowledge first epistemology 64
Kunda, Ziva 91
Kusch, Martin 11

Lackey, Jennifer 21n.2, 129–31, 169
liberatory epistemology 9, 135–8
    (*see also* feminist epistemology; social
    situation; virtue epistemology)
liberal attitudes towards science 53–4, 66–8,
    159–61 (*see also* conservative attitudes
    towards science; motivated reasoning;
    political ideology)
Levy, Neil 78n.4, 166
Lodge, Milton 97, 158–9
Lord, Charles G. 96–7

McKenna, Robin 13
Medina, José 138–42, 147–53, 155–6
Meehan, Daniella 78–81
McGlynn, Aidan 138
Mill, John Stuart 16–17, 106–10, 114, 120–4

Mills, Charles 5–6, 19, 24–8, 34–5, 39–40,
    43, 138
misinformation 50, 54–5, 68–70
Moser, Susanne C. 67
motivated reasoning 54–5, 68–9, 96–9, 158–63,
    165 (*see also* conservative attitudes towards
    science; liberal attitudes towards science;
    political ideology)

naturalized epistemology 36, 165, 177
normative theory 26, 35–6
norms of inquiry 14–15, 36–7, 105n.1
    (*see also* inquiry epistemology)
nudges 72–5, 80–1, 93, 95–6 (*see also* beneficent
    nudge programme, the)

open-mindedness 125–6, 136–7, 139
    (*see also* intellectual virtues; virtue
    epistemology)
objectivity 9–10
oppression 25–6, 136, 139–40 (*see also* epistemic
    oppression)

Pasnau, Robert 40n.4, 177–8
paternalism 60, 71–3, 81–2, 100–1
    (*see also* autonomy; epistemic
    paternalism; nudges)
persuasion 62–3, 73–4, 81–4
Pew Research Center 53–4, 67
Piovarchy Adam 145–6
philosophical methodology 13–14, 177–8
politically motivated reasoning (*see* motivated
    reasoning)
political epistemology 33
political ideology 67–8, 96–7, 163n.2
    (*see also* motivated reasoning)
prebunking 69–70, 103 (*see also* science
    marketing)
prejudice 118, 122–3, 140–1, 144–6, 154
    (*see also* epistemic injustice; responsibility)
problem of the second best, the (*see* fallacy of
    approximation, the)
propositional vs. doxastic justification 167–8
Pritchard, Duncan 129n.5
public ignorance 67, 161–2
public policy 50–1, 62

Quine, Willard van Orman 36

rationality 109, 177
Rawls, John 19, 22–3, 27–8, 37
realistic vs. utopian theory 23–4, 29–31
    (*see also* idealization; ideal vs. non-ideal
    theory; ideal epistemology)

reasons, responsiveness to 114–15
regulative epistemology 14n.3, 90–1
regulative ideals 10
reliabilism 134, 163–5, 172n.4 (*see also* epistemic
   externalism)
Rescher, Nicholas 10n.2
responsibility
   accountability vs. attributability
     responsibility 146
   forward-looking vs. backward-looking
     conceptions of responsibility 144, 147
   responsibility and blame 115, 143–4,
     146
   responsibilities as inquirers 135–6, 141–2
   responsibility in non-ideal epistemology 8–9
   shared responsibility 142
   social–connection model of responsibility,
     the 152–3
   (*see also* active ignorance; blame; epistemic
     injustice; epistemic obligations; epistemic
     responsibility; prejudice; white ignorance)
Riley, Evan 74–8, 80–1
Roberts, Robert C. 91, 98–100, 103, 134–5

scepticism
   old vs. new scepticism 177
   scepticism about the external world 30, 59
   scepticism about our scientific beliefs 128–9,
     161, 164, 174–6
   scepticism and the first-person
     perspective 175
   scepticism in non-ideal epistemology 177–9
science communication 7, 69–70, 84
science denial
   global warming 52–4, 62, 66–7, 159–61
   nuclear waste disposal 97–8, 159–63
   vaccines 38, 78–9, 163n.2
science marketing 66, 69–70, 71n.2, 72, 74, 76–7,
    81–2, 84, 102–3 (*see also* framing;
    prebunking; spokesperson, the choice of)
scientific literacy 67–9
silence 129–30
silencing 118 (*see also* testimonial quieting;
   testimonial smothering)
Simion, Mona 36
social and political values 9–10, 68, 98–9, 137,
   168–9
social epistemology 2–4, 6–8, 12, 21, 31–7, 40n.4,
   57–8
social identity (*see* social situation)
social imaginary, the 140–2
social institutions 6–8, 12–13, 25–6, 33, 55–6, 64
   (*see also* idealization; institutional
   epistemology)
social interaction 2–3, 5, 11–12, 21, 31–2, 131–3
   (*see also* engagement; idealization)

social power 5–6, 32, 130, 136–7 (*see also*
   engagement; idealization; social interaction)
social roles (*see* social situation)
social segregation 53
social situation 6–9, 13, 26, 31–2, 49, 117, 131,
   135–6, 139, 150–2 (*see also* feminist
   epistemology; liberatory epistemology)
social structures 138–40, 152
sociology of scientific knowledge 11
Sosa, Ernest 134, 137
spokesperson, the choice of 69 (*see also* science
   marketing)
Srinivasan, Amia 124–5
stereotypes (*see* prejudices)
Sunstein, Cass R. 72–3, 75, 93–4

Taber, Charles S. 97, 158–9
Tanesini, Alessandra 139, 143
Tessman, Lisa 136–7
testimonial quieting 118–19 (*see also* epistemic
   oppression; testimonial smothering)
testimonial smothering 118–19 (*see also* epistemic
   oppression; testimonial quieting)
Thaler, Richard H. 72–3, 75, 93–4
traditional epistemology 2–4, 6, 21, 31–3
   (*see also* virtue epistemology)
trust and trustworthiness 51
Tsai, George 82–4
twentieth-century epistemology 21, 40n.4

US National Academy of Sciences 160

Valentini, Laura 21–4, 30–1
van der Linden, Sander 68
Vavova, Katia 174–5
veritism 64
vice epistemology 133n.1, 143
   (*see also* intellectual vices; responsibility)
virtue epistemology 32, 134
   traditional vs. liberatory virtue
     epistemology 133–6
   (*see also* intellectual virtues)

Watson, Gary 146
Weisberg, Michael 39–40
Welch, Brynn 73–4
well-ordered science 64
white ignorance 34, 138 (*see also* active
   ignorance; responsibility)
White supremacy 138
Williamson, Timothy 64
Wood, W. Jay 91, 98–100, 103, 134–5

Young, Iris M. 152

Zagzebski, Linda 134